T3-BRK-859

MTBE:
Effects on Soil and Groundwater Resources

MTBE:
Effects on Soil and Groundwater Resource

Edited by

James Jacobs
Jacques Guertin
Christy Herron

363.738
M939

Library of Congress Cataloging-in-Publication Data

MTBE : effects on soil and groundwater resources / James Jacobs, Jacques Guertin, Christy Herron.
p. cm.
Includes bibliographical references and index.
ISBN 1-56670-553-3 (alk. paper)
1. Petroleum as fuel—Additives—Environmental aspects. 2. Butyl methyl ether—Environmental aspects. 3. Groundwater—Pollution. 4. Soil pollution. 5. Butyl methyl ether—Toxicology. I. Jacobs, James J., 1943– II. Guertin, Jacques. III. Heron, Christy.
TD427.P4 M83 2000
363.738—dc21 00-058780
CIP

This book contains information obtained from authentic and highly regarded sources. Reprinted material is quoted with permission, and sources are indicated. A wide variety of references are listed. Reasonable efforts have been made to publish reliable data and information, but the author and the publisher cannot assume responsibility for the validity of all materials or for the consequences of their use.

Neither this book nor any part may be reproduced or transmitted in any form or by any means, electronic or mechanical, including photocopying, microfilming, and recording, or by any information storage and retrieval system, without prior permission in writing from the publisher.

The consent of CRC Press LLC does not extend to copying for general distribution, for promotion, for creating new works, or for resale. Specific permission must be obtained in writing from CRC Press LLC for such copying.

Direct all inquiries to CRC Press LLC, 2000 Corporate Blvd., N.W., Boca Raton, Florida 33431.

Trademark Notice: Product or corporate names may be trademarks or registered trademarks, and are used only for identification and explanation, without intent to infringe.

Visit the CRC Press Web site at www.crcpress.com

© 2001 by CRC Press LLC

Lewis Publishers is an imprint of CRC Press LLC

No claim to original U.S. Government works
International Standard Book Number 1-56670-553-3
Library of Congress Card Number 00-058780
Printed in the United States of America 2 3 4 5 6 7 8 9 0
Printed on acid-free paper

Preface

The nationwide controversy over MTBE had reached such a level that, while on a geology field trip in June 1997 to Salmon Lake in California's Sierra Nevada mountains, I suggested to Bill Motzer that we write an article or series of articles summarizing and evaluating the existing data about MTBE. After discussing the issue with several other technical specialists, the Independent Environmental Technical Evaluation Group (IETEG) was founded in September 1997. After 3 years, the IETEG presents this publication as the result of our collaboration.

Members of the IETEG are Dr. Bill Motzer, Forensic Geochemist (Physical and Chemical Properties of MTBE); Dr. Jacques Guertin, Toxicologist/Chemist (Toxicity, Health Effects, and Taste and Odor Thresholds of MTBE; Appendix I, Toxicity of MTBE: Human Health Risk Calculations); Fred Stanin, Hydrogeologist (Transport and Fate of MTBE in the Environment); Dr. Paul Fahrenthold, Remediation Engineer/Chemist (Detection and Treatment of MTBE in Soil and Groundwater; Appendix G, Synthesis, Properties, and Environmental Fate of MTBE and Oxygenate Chemicals); Markus Niebanck, Hydrogeologist (MTBE: A Perspective on Environmental Policy); Christy Herron, Environmental Planning and Policy (Introduction, History and Overview of Fuel Oxygenates and MTBE, MTBE: A Perspective on Environmental Policy); David Abbott, Geologist (References and Reading List, technical review); Russell Pfeil, Geologist (project compilation), and James Jacobs, Hydrogeologist (group and publication leader; Introduction; History and Overview of Fuel Oxygenates and MTBE; Conclusions and Recommendations; Appendix F, MTBE: Subsurface Investigation and Clean-Up; Appendix H, Plume Geometries for Subsurface Concentrations of MTBE). Appendix E, Geologic Principles and MTBE, was contributed by Stephen Testa and appears in the previously published *Geological Aspects of Hazardous Waste Management* (1993); this section has been updated somewhat since its previous publication. The IETEG is grateful for the review of and contribution to the manuscript-in-progress by Clifton Davenport, Hydrogeologist; Dr. Brendan Dooher, Research Engineer; Dr. Angus McGrath, Geochemist; and Stephen Testa, Geologist.

University Libraries
Carnegie Mellon University
Pittsburgh, PA 15213-3890

The IETEG is also grateful to Christy Herron for compilation and editing of sections of this book; to Bonnie Dash for technical review of the manuscript; and to James Geluz for publication design, illustration, and layout. The IETEG also thanks Steve Testa for his suggestions on publishing as well as his written contributions. The authors are deeply grateful to our respective families and friends, without whose understanding, support, and encouragement this project could never have been completed.

James A. Jacobs
President & Co-founder
IETEG

William E. Motzer
Co-founder
IETEG

May 2000

About the authors

David Abbott, B.S., Geology

David Abbott, B.S. in geology, has more than 20 years of experience in the development of water supplies, protection of groundwater resources, and perennial yield investigations throughout the western United States. At David Keith Todd Associates, he has conducted numerous groundwater resource assessments and hydrogeologic studies. He is a registered geologist and certified hydrogeologist in the State of California. Mr. Abbott is a director of the Groundwater Resources Association of California.

Paul Fahrenthold, Ph.D., Remediation Engineer

Paul Fahrenthold, Ph.D. in chemistry, specializes in process engineering and environmental remediation and cleanup. He has conducted numerous projects involving recovery of raw materials and treatment of groundwater, storm water, and soil at contaminated sites, including determining the extent of contamination using extensive sampling and chemical analysis methods. Dr. Fahrenthold also conducts short courses and workshops on environmental chemistry and forensic geochemistry. For the last 33 years, he has worked for a number of environmental consulting firms as well as the U.S. EPA. Currently, Dr. Fahrenthold is president of his own consulting firm, Fahrenthold & Associates, Inc., in Concord, California, and is CEO of OXXL Corporation.

Jacques Guertin, Ph.D., Toxicologist-Chemist

Jacques Guertin, Ph.D. in chemistry, has more than 20 years of experience in environmental science. He holds 5 U.S. patents and is author of more than 70 technical publications. He specializes in toxicology, health-ecological risk assessment, and computer hardware-software, and is an expert in sampling and chemical analysis and materials science. He has worked at Bell Telephone Laboratories, the Electric Power Research Institute, and environmental consulting firms. Currently, Dr. Guertin has his own environmental consulting business in Newark, California. In addition, he teaches environmental science, risk assessment, forensic science, chemistry, and materials science at the University of California and University of Wisconsin Extension, as well as advanced chemistry (college preparation), physics, earth science, and astronomy to high school students.

Christy Herron, B.A., Environmental Planning and Policy
Christy Herron, B.A. in environmental studies and English, has 5 years of environmental consulting experience. Ms. Herron specializes in public policy, and environmental planning and land use issues. Ms. Herron has also written on other subjects of interest to the environmental industry, including geothermal and renewable energy in the deregulated California energy market.

James Jacobs, M.A., Hydrogeology and Subsurface Investigation
James Jacobs, M.A. in geology, has more than 20 years of experience in subsurface geology. He specializes in in-situ remediation delivery systems and soil and groundwater drilling methods. He is registered as a geologist in several states and is a certified hydrogeologist in the State of California. Mr. Jacobs is president of FAST-TEK Engineering Support Services. He is also past president of the Groundwater Resources Association, San Francisco Chapter, and the California Section of the American Institute of Professional Geologists. He is currently an incorporator and director of the California Council of Geosciences Organizations and a director of the Groundwater Resources Association of California. Mr. Jacobs is also leader and co-founder of the IETEG.

William E. Motzer, Ph.D., Forensic Geochemist
Bill Motzer, Ph.D. in geology, has more than 21 years of experience in geology, exploration geochemistry, environmental geology, environmental forensics, and forensic geochemistry. During the past 14 years, he has conducted over 350 surface and subsurface environmental investigations and site remediation programs in California, Arizona, Oregon, Idaho, Nevada, Utah, Washington, and Colorado. He has provided expert witness testimony in forensic geochemistry and environmental forensics. Dr. Motzer has also taught courses in applied environmental geochemistry and forensic geochemistry for the University of California, Berkeley Extension and the University of Wisconsin, Madison Extension. Currently, he is Western Regional Manager at Hydro-Environmental Technologies, Inc. in Alameda, California. He is a registered geologist in six states, a California registered environmental assessor, and a certified professional geologist with the American Institute of Professional Geologists. He is also co-founder of the IETEG.

Markus Niebanck, M.S., Hydrogeologist
Markus Niebanck, M.S. in geology, is a registered geologist in several western states, including California. He has 14 years experience in the environmental industry, especially with fuel hydrocarbon contamination. Mr. Niebanck is a founding partner of Clearwater Group, Inc., in Oakland, California.

Frederick Stanin, M.S., Hydrogeologist

Fred Stanin, M.S. in geology, has more than 20 years of professional experience. He has conducted numerous in-situ remedial tests for soil and groundwater contaminated with fuel hydrocarbons, often using risk-based corrective action (RBCA) strategies. He is a registered geologist and certified hydrogeologist in the State of California. Currently, Mr. Stanin is with Parsons Engineering Science in Oakland, California, conducting environmental restoration at the Ernest Orlando Lawrence Berkeley National Laboratory.

Contents

chapter one

Introduction

> The significant problems we face cannot be solved at the same level of thinking we were at when we created them.
> Albert Einstein
> (1879-1955)

1.1 Purpose

The Independent Environmental Technical Evaluation Group (IETEG) — a research organization located in Point Richmond, California — was created in 1997 to present objective scientific and engineering information about controversial environmental issues as a foundation for rational discussion and policy development. The IETEG, comprising members of the environmental consulting industry and academia, are specialists in the assessment, mapping, and remediation of soil and groundwater, as well as environmental compliance and environmental risk assessment. Members of the group review existing data and articles in as unbiased a manner as possible, with the purpose of developing expert, objective viewpoints about controversial issues.

As scientists and engineers, the authors of this publication are primarily concerned with human ingestion of MTBE through drinking contaminated groundwater or surface water. Dermal contact through bathing or washing in MTBE-contaminated water is another exposure pathway. Inhalation by the general public, although an important mode of human exposure, is less likely to be associated with exposure to MTBE-contaminated soils and groundwater.

This publication focuses primarily on MTBE contamination in groundwater; contamination of surface water is also discussed.

Finally, this publication focuses on MTBE's health risk only as a secondary concern in comparison to the potential that MTBE has to affect the taste and odor of water from public wells, and to degrade the beneficial uses of groundwater.

1.2 Introduction to the MTBE problem

Methyl *tertiary*-butyl ether (MTBE) is a synthetic compound that was developed as a technological solution to a technology-derived problem created by air pollution from vehicle emissions. MTBE was added to gasoline with the intent to reduce air emissions, making the fuel burn cleaner. Ironically, use of this air-saving gasoline additive has created one of the most threatening and widespread environmental problems for the nation's drinking water supply. MTBE is highly soluble in water, more than 75 times more so than many other gasoline compounds. As of the date of this publication, MTBE has been found in groundwater at over 250,000 contaminated sites throughout the nation.

MTBE remains at the center of a nationwide debate. Some parties claim that the use of MTBE in reformulated gasoline (RFG) allows refiners to produce cleaner-burning fuels, reducing carbon monoxide (CO) concentrations in the air. Others claim exposure to MTBE causes illness. Scientific experts have lined up on both sides of the debate. Because conclusive data on the health effects as well as beneficial qualities of MTBE were (and, as of the date of this publication, still are) lacking, the tenor of the discussion has sometimes been quite strange, with both sides using the same limited bits of scientific data to support opposing viewpoints.

Many of MTBE's supporters changed their positions during the 1990s. One by one, oil companies that were once its most staunch defenders began denouncing MTBE. This dramatic change took place after oil producers calculated and fully realized the cost of removing MTBE from groundwater. The Western States Petroleum Association (WSPA)[1] was also at one time a vocal proponent of using MTBE in RFG to reduce CO emissions; WSPA, too, eventually changed its position to one in support of alternatives to MTBE.

In July 1999, the United States Environmental Protection Agency (U.S. EPA) concluded a blue-ribbon panel study of the use of MTBE in gasoline. The U.S. EPA had previously supported the use of MTBE to achieve air quality goals; however, after the conclusion of its study, the U.S. EPA recommended discontinuing the use of MTBE because of the implications for contamination of groundwater resources.

By now, the full extent of MTBE's impact to groundwater resources has been acknowledged by regulators, oil companies, and the general public alike.

During the late 1990s, the nationwide drama surrounding MTBE played itself out more visibly in some states and regions than in others, often dependent on the extent to which these states use groundwater as a source of drinking water. In California, for instance, opposing viewpoints about MTBE voiced by oil refiners, citizen activist groups, air quality specialists, and water quality specialists often make front-page news. Public outcry and concerns over the unknown health effects of MTBE and the cost of cleanup continue to accumulate.

1.2.1 MTBE and air quality: the continuing debate

The Oxygenated Fuels Association (OFA), not surprisingly, defended and continues to defend the use of MTBE. OFA argues the following in their publication "Gasoline Reformulated with Methyl *Tertiary*-Butyl Ether (MTBE):"[2]

- "Air quality has markedly improved over the last 20 years."
- "MTBE is not hazardous to health under the conditions of intended use."
- "Health effects from exposure to MTBE while refueling or driving . . . are rare, if they exist at all."
- "Well-conducted scientific studies have not demonstrated any causal relationship between human health complaints and use of MTBE-containing RFG."

OFA's tone and viewpoint have not changed recently regarding the use of MTBE in reformulated gasoline. In February 2000, in response to public calls for the elimination of the federally mandated oxygenate standard, the OFA stated that it believed the elimination of the oxygenate standard could lead to "a national fuel disaster which would result from the uncertainty of untested alternative fuel sources and their implications on health, safety, cost, and transportation concerns...." The OFA also stated, "Despite its key contribution to the nation's enormously successful clean-burning gasoline programs, MTBE has been unfairly singled out as a threat to groundwater."[3]

California's Air Resources Board (CARB) has maintained that the use of MTBE in gasoline is, on balance, ultimately useful in order to achieve air quality goals, even though air quality benefits are uncertain, under debate, and increasingly discounted. Determining whether MTBE usage has resulted in a decline in urban concentrations of airborne pollutants is complicated by the overall decline in all such pollutants. This decline can in no small part be attributed to more stringent U.S. EPA emissions standards, along with an improvement in automobile technology for emission control.[4]

The air quality division of the U.S. EPA has supported MTBE as a beneficial tool in reducing automobile emissions. The model that the U.S. EPA has used to estimate the benefits of oxygenated gasoline on carbon monoxide (CO) emissions, however, probably overestimates these benefits by approximately a factor of 2.[4]

1.2.2 MTBE and groundwater quality: no debate here

MTBE in groundwater can originate from point and non-point sources. Potential point sources of MTBE include leaking underground gasoline storage tanks, pipelines, and gasoline spills. Leaking underground storage tanks (USTs) containing gasoline are the major source of MTBE contamination. About 22% of the 1.2 million USTs at more than one-half million cleanup sites in the country had leaked as of July 1994.[5]

From California to Maine, scientists and legislators have expressed more and more concern about the potential for MTBE to contaminate groundwater. In 1995, a letter written to Maine's governor by a state representative stated: "This will be a most costly expenditure to the State if we have to clean up our drinking water supplies because MTBE contaminated our lakes and private wells."[6]

A feature story about the national MTBE controversy aired on the CBS television show "60 Minutes" on January 16, 2000. On January 17, on the heels of this news story concerning the widespread groundwater contamination caused by MTBE, the Board of Directors of the Groundwater Resources Association of California (GRA) adopted a resolution renewing its request for an immediate nationwide ban on MTBE.

At the meeting of the Board of Directors, GRA Board President Timothy Parker warned that

> "...widespread MTBE contamination appears to be of catastrophic proportions. The problem is symptomatic of our failure to understand the full magnitude of the health risks associated with commercial chemicals before they are introduced into the environment.
>
> "When one considers the level of review provided to chemicals that are ingested in food before they are authorized by the Food and Drug Administration, it is amazing how little consideration is given to something with such an obvious potential to contaminate our water supply."[7]

Although public opinion has become weighted against the use of MTBE in gasoline, the debate over the relative merits and drawbacks of MTBE is ongoing. From the scientific community to the public at large, however, those who argue the beneficial vs. detrimental effects of MTBE generally agree that the cumulative threat MTBE poses to groundwater needs to be addressed.

On March 26, 1999, California Governor Gray Davis instituted a 4-year phase-out of MTBE in gasoline in that state. Federal action recommending a decrease in the use or elimination of MTBE in gasoline constitutes the most recent judgment passed on MTBE, as of the date of this publication.

1.3 Some facts about MTBE

Some facts about the nature of MTBE and MTBE contamination of groundwater in the U.S. can be stated clearly, and are listed below.

- TASTE AND ODOR: Water sources contaminated with very low levels of MTBE become unusable for human consumption because of the unpleasant, turpentine-like taste and odor of MTBE.

- HEALTH RISK: Although MTBE is considered a potential health risk, there is little to no evidence that MTBE causes cancer in humans. Fortunately, the strong, turpentine-like taste and odor of MTBE can be detected by humans at relatively low concentrations in water. The potential for the population at large to drink significant quantities of water highly contaminated with MTBE is therefore unlikely. MTBE is not listed as a human carcinogen by the U.S. National Toxicology Panel, the California Proposition 65 Committee, or the International Agency for Research on Cancer.
- FATE AND TRANSPORT: MTBE enters soil and groundwater systems through leaking USTs, surface spills of gasoline, and other sources. Due to its high solubility in water, MTBE tends to migrate much faster and further in groundwater than equal amounts of other gasoline compounds.
- REMEDIATION: Owing to the physical and chemical characteristics of MTBE, remediation (or cleanup) of MTBE in groundwater is expensive, time consuming, and technically challenging. New technologies might improve remediation efficiency and reduce cost.
- AIR POLLUTION: The source of the reductions in air pollution that have taken place over the last few years is still under debate. Part of the overall nationwide improvement in air quality can be attributed to newer and more efficient automobile engines, which produce fewer harmful emissions. Reformulated gasoline containing MTBE may be responsible for some of the improvement in air quality, as suggested by the U.S. EPA; however, there is some disagreement among air quality studies as to how much the use of MTBE in gasoline reduces automobile air emissions.
- ALTERNATIVE OXYGENATES: If other oxygenates are added to gasoline in the place of MTBE to lower vehicle emissions, the cost and currently limited availability of these alternatives, such as ethanol, are likely to increase the cost of gasoline. In addition, replacements for MTBE must be evaluated carefully for their potential health effects and their fate and transport characteristics in the subsurface.

1.4 How to read this publication

This publication comprises eight chapters, some of which present more technically specialized material than others.

1. Introduction
2. History and Overview of Fuel Oxygenates and MTBE
3. Physical and Chemical Properties of MTBE
4. Toxicity, Health Effects, and Taste and Odor Thresholds of MTBE
5. Transport and Fate of MTBE in the Environment
6. Detection and Treatment of MTBE in Soil and Groundwater

7. MTBE: A Perspective on Environmental Policy
8. Conclusions and Recommendations.

A glossary of terms used, a series of technical appendices, a bibliography and reading list, and an index are also presented at the end of this publication.

Although the scope of this publication is intended to be national, many chapters draw on studies of MTBE contamination in groundwater in California, a state plagued with some of the most severely MTBE-contaminated groundwater in the nation.

Endnotes and references

[1] The Western States Petroleum Association (WSPA) is a non-profit trade association representing approximately 36 companies that account for most petroleum exploration, production, refining, transportation, and marketing activities in six western states: Arizona, California, Hawaii, Nevada, Oregon, and Washington. (Source: www.wspa.org/aboutus.htm)

[2] Oxygenated Fuels Association, Gasoline Reformulated with Methyl *Tertiary*-Butyl Ether (MTBE), Arlington, Virginia, April 1996.

[3] Oxygenated Fuels Association, Statement of the Oxygenated Fuels Association (OFA) in Response to Calls for the Elimination of the Oxygenated Standard in Reformulated Gasoline, www.ofa.net/NESCAUM-ALA-APIResponseStatement.html, February 3, 2000.

[4] National Science and Technology Council, Interagency Assessment of Oxygenated Fuels, Committee on Environmental and Natural Resources, Executive Office of the President of the United States, June 1997.

[5] U.S. EPA, UST Program Facts—Cleaning Up Releases: EPA 510-F-94-006, Office of Solid Waste and Emergency Response, August, 1994.

[6] Lovett, G.P., Letter from Maine Representative Glenys P. Lovett, 21st District, to Governor Angus King, August 4, 1995.

[7] Groundwater Resources Association, Groundwater Resources Association Applauds "60 Minutes" Story on MTBE Contamination, *Hydrovisions*, v. 8, n. 4, Winter-Spring 1999–2000.

chapter two

History and overview of fuel oxygenates and MTBE

2.1 *The origin of gasoline additives*

Ever since the early days of the automobile, petroleum refiners have worked to increase the combustion efficiency of their product, usually by the addition of octane-enhancing fuel additives. Industrial ethanol, traditionally manufactured by the fermentation of plant material, is one such fuel additive, though it has been plagued through the years by its popular association with beverage ethanol or whiskey. Industrial ethanol, for instance, was taxed for some time in exactly the same manner as beverage alcohol. The moral taint associated with whiskey production extended to industrial ethanol most notably during Prohibition. To this day, federal law requires the denaturing (the addition of a small amount of a poisonous substance) of industrial ethanol to prevent its consumption.

Organic lead, another octane-enhancing gasoline additive, eventually became the additive of choice for refiners. Lead was less "bulky" than ethanol — in other words, it took up less space in the gas tank. Lead did not suffer, as ethanol did, from an association with an external moral issue — until the 1970s, that is, when lead's detrimental environmental effects became widely recognized and denounced. The public outcry over these effects, coupled with the discovery of lead's damaging effects on emission control devices, resulted in the phase-out of the use of leaded gasoline in California, followed by a federal phase-out.

Ethers, such as methyl *tertiary*-butyl ether (MTBE), replaced lead as some of the petroleum industry's additives of choice. The continuing quest for a better gasoline additive, however, did not end with the introduction of ethers. Recent health studies and public complaints have implicated MTBE as a possible human health hazard, especially through the inhalation of MTBE fumes or vapors. In addition, MTBE contamination has become a serious threat to groundwater resources. MTBE seems to be subject to the same curious blend of scientific study and public denouncement as its predecessors.

2.1.1 Oxygenates as gasoline additives

Oxygenated gasoline is designed to increase the combustion efficiency of fuel, thereby reducing carbon monoxide (CO) emissions. Oxygenates have been used in gasoline in the U.S. since the 1930s, when alcohols were added to gasoline to enhance octane. By the 1950s, a publication by the American Petroleum Institute referenced the potential for adding MTBE to gasoline. During the oil shortages of the mid-1970s, ether-based compounds such as MTBE were added to gasoline to extend the use of the gasoline and as octane enhancers. By 1978, the Gasohol Program in the U.S. began using a gasoline blended with ethanol (10% ethanol by volume). By the 1980s, stringent phase-out requirements for lead in gasoline resulted in a further increase in the addition of ethers to gasoline. By the late 1980s, MTBE was blended into gasoline sold nationwide to meet federal requirements for reformulated gasoline; by the late 1990s, widespread MTBE contamination in groundwater had resulted in a nationwide crisis.[1]

The first winter oxygenated gasoline program in the nation was implemented in Denver, Colorado in 1988. During the 1980s, oxygenates began to be used more widely as some states implemented oxygenated gasoline programs for the control of CO during cold weather.[2] The 1990 Amendments to the federal Clean Air Act (1990 CAA) require at least a 2.7% oxygen content for gasoline sold in CO nonattainment areas. This level of oxygen is typically achieved by the addition of an oxygenate. In CO nonattainment areas, the 2.7% oxygen content for gasoline has typically been achieved by the addition of about 15% MTBE[3] or about 7.5% ethanol by volume. The use of MTBE is not specifically mandated by the 1990 CAA, but MTBE tends to be the additive of choice for most gasoline vendors. Other fuel oxygenates that are in use to a lesser extent include the following:

- *Tertiary*-butyl alcohol (TBA)
- Di-isopropyl ether (DIPE)
- Ethyl *tertiary*-butyl ether (ETBE)
- *Tertiary*-amyl methyl ether (TAME)

Methanol, ethanol, and TBA are alcohols, while MTBE, DIPE, ETBE, and TAME are ethers.

TBA has been found at concentrations of 1,100 micrograms per liter, or parts per billion (ppb) in groundwater at a gasoline service station site in San Joaquin County in August 1997, suggesting that TBA has also been used in gasoline in California. DIPE is primarily in use on the East Coast. According to California's Central Regional Water Quality Control Board, as of August 30, 1997, there was no information regarding the use of ETBE in California.[4] TAME has been added to California fuels since 1995, and has been found in groundwater in Southern California and in San Joaquin County.

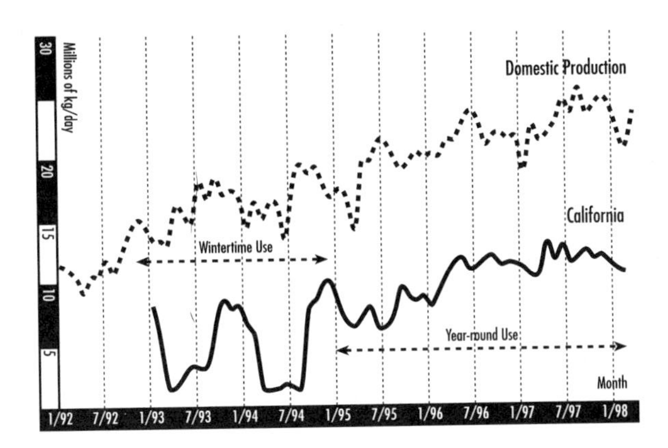

Figure 2.1 Total domestic production and California consumption in terms of total refinery inputs of MTBE. The transition from the wintertime to the year-round program in California is clearly evident (10^6kg/day = 2.8 · 10^6 gal./day^{-1}). (Data from the Office of Oil and Gas, Energy Information Administration, U.S. Department of Energy) (From An Evaluation of MTBE Impacts to CA Groundwater sources, LLNL, Happel et al., 1998. With permission.)

MTBE has been the most common fuel oxygenate, used in more than 80% of oxygenated fuels.[5] Figure 2.1 shows the increase in MTBE use in California from 1993 to 1998.

2.2 MTBE — benefits and perceived costs

MTBE and other oxygenates serve a dual purpose: they increase the combustion efficiency of gasoline, and also reduce the amount of harmful emissions, such as CO and ozone, that are the direct or indirect result of incomplete automobile combustion. Some cities and regions, such as Denver, Colorado, already had oxygenated fuel or RFG programs in place before the 1990 CAA took effect; many programs began in other regions in the winter of 1992 to 1993.[6]

2.2.1 MTBE in air

Shortly after these programs were implemented broadly in the U.S., there were widespread public health complaints of nausea, dizziness, and respiratory complications. No studies conclusively proving that MTBE was a human inhalation health threat had been conducted before its widespread use as a gasoline oxygenate. Despite this fact, public complaints that various respiratory and nervous health complaints were linked to the inhalation of airborne MTBE began increasing daily.

The main human health concerns related to MTBE were initially associated with exposure to airborne MTBE. Health complaints were reported in

Fairbanks, Alaska in 1992, when 200 residents reported dizziness, irritated eyes, burning of the nose and throat, coughing, disorientation, and nausea after MTBE had been added to gasoline in that state. In 1994, the American Medical Association (AMA) issued a resolution addressing MTBE. This resolution was approved at the June 14 annual meeting of the AMA:

> Whereas, in Fairbanks and Anchorage in 1992–1993, a large number of citizens complained of symptoms including headaches, dizziness, nausea, cough and eye irritation; and studies by the Alaska Division of Public Health and the National Centers for Disease Control and Prevention found that these symptoms were associated with exposure to oxygenated gasoline, that MTBE was detectable in the blood of all workers and communities studied in Fairbanks ...
> The AMA urges that a moratorium on the use of Methyl *tertiary*-butyl ether (MTBE) blended fuels be put into place until such time that scientific studies show that MTBE fuels are not harmful to health, and that no penalties or sanctions be imposed on Alaska during the moratorium.[7]

Even though more and more health studies continued to debunk theories that exposure to low levels of MTBE, as experienced by the general public (including motorists), does not result in chronic aggravation of the respiratory or central nervous systems, the number of anecdotal reports and health complaints linking MTBE to these symptoms continued to increase. The suggestion that these health effects are purely anecdotal, stimulated or created by negative publicity, or even, as hypothesized in one study, psychogenic in nature[8] does little to dismiss them. Whether purely anecdotal or not, public health complaints about MTBE are widespread and not confined to any one region or state. This public reaction has had a direct influence on many states' air quality policies addressing MTBE. In addition, the perception of MTBE as an inhalation health risk cannot help but affect another environmental policy, such as that regulating MTBE in groundwater.

In addition to health complaints, there have been complaints of reduced fuel economy and engine performance from the use of MTBE in gasoline.[2]

2.2.2 MTBE in groundwater

The most emphatic public outcry over MTBE had, until recently, been that regarding inhalation health risks. A recently realized environmental impact of MTBE — its potential hazard to the beneficial use of water supplies — has arisen which has been even more influential in the final determination of MTBE policy.

MTBE has been the most common fuel oxygenate, used in more than 80% of oxygenated fuels.[5] Figure 2.1 shows the increase in MTBE use in California from 1993 to 1998.

An informal study conducted by the Santa Clara Valley Water District (SCVWD) in California reported that MTBE was detected (tasted) by three of four test subjects at concentrations as low as 10 ppb.[9] Other published research suggests that the concentrations at which human ingestion of MTBE creates a cancer or other health risk are likely to be far higher than these levels.[6] This is reflected in the U.S. Environmental Protection Agency (U.S. EPA)'s December 1997 Drinking Water Advisory for MTBE, subtitled "Consumer Acceptability Advice and Health Effects Analysis,"[10] in which the concentration of MTBE in drinking water that the U.S. EPA has determined to be acceptable to most consumers is set between 20 and 40 ppb. The U.S. EPA advisory states that exposure levels resulting in cancer or noncancer effects in rodent tests are 20,000 to 100,000 times higher than this range; additionally, no tests conclusively linking any levels of ingested MTBE with a human cancer or noncancer health risk have been conducted as yet. It should be noted that studies with animals have shown a correlation between MTBE ingestion and occurrence of cancer.[6]

The degradation of water supplies by MTBE contamination, specifically with regard to taste and odor considerations, has already seriously affected numerous public wellfields across the U.S. In 1997, the City of Santa Monica, California shut down half of its water wells because of MTBE contamination, suffering a 75% loss of the local groundwater supply; the city spent $3 million importing water for its use.[11] More recently, water supply wells have been affected in South Lake Tahoe, California by MTBE contamination; also, at least one water company in Santa Clara County, California has shut down a public well because well water was found to be contaminated with MTBE.

On February 17, 1998, the SCVWD board voted to send a letter to Governor Wilson urging the removal of MTBE from gasoline. This water district has taken the position that MTBE in gasoline, especially gasoline contaminating groundwater supplies from leaking gasoline underground storage tanks, poses a serious threat to groundwater resources.[12]

In 1997, a MTBE Media Fact Sheet from the Santa Clara Valley Water District (SCVWD) in California states:

> "Increasingly, MTBE is finding its way into groundwater, in storm runoff, and in some cases, drinking water wells and underground aquifers. Once in groundwater, it persists there and is expensive to remove."[13]

In Maine, the presence of MTBE and other gasoline components in groundwater was evaluated in a study issued in 1998 by the State Department of Environmental Protection. Water samples were collected from 951 randomly selected household wells and other household water supplies such as springs and lakes. MTBE was detected in 150, or 15.8%, of the 951 private wells sampled. 1.1% of the water samples showed levels of MTBE above the Maine drinking water standard of 35 ppb. If extrapolated statewide, these numbers suggest that these levels of MTBE were present in 1400 to 5200 private wells in the state. In comparison, other gasoline compounds were infrequently detected above Maine's heath-based standards.[14]

The United States Geological Survey (USGS), as part of their National Water-Quality Assessment Program, collected samples of shallow groundwater from wells located in 8 urban and 20 agricultural areas in the U.S. from 1993 to 1994. In this survey, MTBE was one of the two most frequently detected volatile organic compounds. MTBE was detected in 27% of urban wells tested — in Denver, Colorado, 79% of groundwater samples had detectable concentrations of MTBE, and in New England, 37% of the samples taken had detectable concentrations. The USGS concluded from the data compiled in this study that MTBE tends to occur most often in shallow groundwater underlying urban areas.[15] Based on this study, it is not unreasonable to assume that the groundwater quality of shallow aquifers in urban areas across the U.S. are threatened with MTBE contamination. Thus, a situation similar to that which shut down half of Santa Monica's wells may very well arise elsewhere.

In their report, "An Evaluation of MTBE Impacts to California Groundwater Resources," released in June 1998, Lawrence Livermore National Laboratory (LLNL) presented several conclusions regarding the potential of MTBE to pose a risk to California's groundwater supplies. Among its other conclusions, the report stated that MTBE is a frequent and widespread contaminant in shallow groundwater throughout California, that it moves relatively quickly through groundwater, and that it is difficult to remove from the groundwater. Based on these conclusions, the LLNL report recommended that, while future research on MTBE is needed, groundwater resources should be managed in order to minimize the potential threat of MTBE.[16]

As of the date of the LLNL report, there were 32,409 leaking UST cleanup sites in California. Of these sites, 13,278 had groundwater that was impacted by gasoline components. Of the sites undergoing active cleanup studied in the LLNL report, 75% were sites impacted by MTBE — bringing the total of MTBE-impacted sites in the state to 10,000. The report also estimated that there are 6700 MTBE-impacted sites in California within 1/2 mile of a drinking water well.[17]

2.3 *The cost of MTBE: environmental headache or public good?*

Given the highly volatile public reaction to the widespread use of MTBE, and given the ambiguous consequences of its use, we see that MTBE occupies a truly complex position as a chemical of environmental concern in relation to the protection of groundwater resources. This has resulted in an unpredictable range of "official positions" taken by members of the petroleum industry and other concerned parties as they scramble to assess probable financial losses from the use of MTBE in gasoline.

Until fairly recently, for instance, several oil companies had supported the widespread use of MTBE as an environmentally safe and economically feasible method of achieving air quality goals. In a 1997 hearing given in Santa Monica by the California Natural Resources Committee, however, a representative of Tosco Corporation (an oil refiner and marketer) presented a letter to the California Air Resources Board outlining Tosco's position as one in favor of reducing or eliminating the use of MTBE in California. Tosco's stated intent was to support California's high air and water quality standards; its position, however, according to the Tosco representative who wrote and presented the letter, is primarily based on the company's belief that the potential investment required to shift from MTBE production and usage would be less than the potential costs of liability for cleanup.[18]

That environmental legislation relating to MTBE cleanup in public water systems will be costly has been evidenced by an environmental cleanup suit, settled in August 1997 in Wilmington, North Carolina. A U.S. District Court awarded $9.5 million to nearly 200 residents of two mobile home parks because MTBE and benzene had contaminated a public drinking-water well. An undisclosed amount of punitive damages was also awarded in this case.

ARCO, the oil company that holds the original patent for MTBE, held an opposite position from Tosco, and supported the use of MTBE. Many or even most other major oil companies, however, have recently taken positions similar to Tosco's, in favor of reducing or eliminating the use of MTBE in gasoline — for example, Chevron, in a December 1997 letter to a California State Senator, stated its support of legislation that would repeal the federal mandate for the use of oxygenated gasoline. The Western States Petroleum Association (WSPA) also supports this legislation.

The controversy regarding MTBE in gasoline in California has also manifested a policy and cost implication with a curious international twist. In June 1999, the Methanex Corporation of Ontario, Canada notified the U.S. government of its intention to seek damages under the North American Free Trade Agreement (NAFTA) relating to California's decision to ban MTBE.[19] Methanex manufactures and markets methanol, and is a major supplier to MTBE producers in the U.S. and elsewhere. The president and CEO of Methanex has stated "The California Governor's Order to ban the use of MTBE in that state unfairly targets MTBE in what are really broader gasoline and water resource issues." Whether Methanex is successful or not in pursuing damages may further affect the future total cost of the use of MTBE in California and the nation.

The discussion about MTBE is not limited to the U.S. Significant contamination of groundwater by MTBE has been found in only three sites in Germany. Unlike the U.S., environmental agencies in Germany have set lower quality (i.e., higher acceptable contamination levels) standards addressing and enforcing underground gasoline storage tank regulations.[20] The underground storage tank upgrades mandated by the U.S. EPA did not require mandatory removal of single-walled underground tanks in the U.S. until December 22, 1998. Due to this relatively late date of required underground tank upgrades, the MTBE problem in the United States is much worse than it might have been had underground tank upgrades such as secondary containment, better monitoring systems, and leak detection alarms been mandated to occur in the late 1980s, when MTBE was initially blended into gasoline on a large scale to meet RFG requirements.

On March 26, 1999, Governor Gray Davis instituted a 4-year phase-out of MTBE in gasoline in California. Federal recommendations to reduce or eliminate the use of MTBE in gasoline in the U.S. have recently, as of the date of this publication, been issued by the U.S. EPA.

In contrast to the phase-out of MTBE in California, which will take several years to implement fully, a recent action by the Texas Natural Resource Conservation Commission (TNRCC) places limits on increases in the use of MTBE, while still retaining what the TNRCC believes to be the air quality benefits of cleaner-burning fuels.

Given the details of its history and the complex public reaction MTBE has inspired, it is not surprising to discover that policy that directly regulates this fuel oxygenate is fractured in its application, differs widely from state to state, and is not always consistent with the U.S. EPA's drinking water advisory for MTBE.

Endnotes and references

[1] Dragos, D., Comparison of the Properties and Behavior of MTBE and the Fuel Oxygenates TAME, ETBE, DIPE, TBA, and Ethanol, National Groundwater Association, San Francisco, March 17, 2000.

[2] National Science and Technology Council, Interagency Assessment of Oxygenated Fuels, Committee on Environmental and Natural Resources, Executive Office of the President of the United States, June 1997.

[3] Vance, D.B., MTBE: Character in Question, *Environmental Technology*, v. 8, n. 1, 1998.

[4] Boggs, G.L. Analysis Required for Oxygenate Compounds Used in California Gasoline–EPA Method 8260 (8240-B and 8020); open file letter, California Regional Water Quality Control Board, Central Valley region, August 30, 1997.

[5] Swain, W., Methyl Tertiary-Butyl Ether File Report, U.S. Geological Survey, ca.water.usgs.gov/mtbe, 2000.

[6] Zogorski, J.S. et al., Fuel Oxygenates and Water Quality: Current Understanding of Sources, Occurrence in Natural Water, Environmental Behavior, Fate and Significance: Interagency Oxygenated Fuel Assessment, Office of Science and Technology, Washington, D.C., 1996.

[7] American Medical Association, Resolution 4-32, A-94, H 135.957, approved at the June 14 annual meeting. "Moratorium of Methyl Tertiary Butyl Ether Use as an Oxygenated Fuel in Alaska," 1994.

[8] Balter, N.J., Acute Health Studies of MTBE, International Center for Toxicology and Medicine Workshop on MTBE/Water Issues, June 4-5, presented by American Methanol Institute, Oxygenated Fuels Association, and Western States Petroleum Association, San Jose, 1997.

[9] Hoose, S., Personal Testimony, Santa Clara Water District Water Resources Management Group, at the (California State) Senate Environmental Quality Oversight Hearing on Leaking Underground Storage Tank Cleanup Program, March 17, 1997. .

[10] U.S. EPA, Drinking Water Advisory, Consumer Acceptability Advice and Health Effects Analysis on Methyl Tertiary-Butyl Ether (MtBE), EPA-822-F-97-009, Office of Water, Washington, D.C., December 1997.

[11] Carlsen, W., Gas Additive's Needless Risk, San Francisco Bay Chronicle, p. A-1, A-8-9, September 15, 1997.

[12] McCabe, M., South Bay Water District Asks for Gas Additive Ban, San Francisco Bay Chronicle, p. A-13, February 18, 1998.

[13] Santa Clara Valley Water District, MTBE Media Fact Sheet, March 31, 1997.

[14] Maine Department of Environmental Protection, MTBE and Other Gasoline Compounds in Maine's Drinking Water Supplies—A Preliminary Report, www.state.me.us/dhs/boh/mtbe/mtbe.pdf, October 13, 1998.

[15] Delzer, G.C. et al., Occurrence of the Gasoline Oxygenate MTBE in Shallow Groundwater in Urban and Agricultural Areas, 1991-95: U.S. Geological Survey Water Resources Investigations Report 96-4145, 1995.

[16] Happel, et al., An Evaluation of MTBE Impacts to California Groundwater Resources, Lawrence Livermore National Laboratory, Environmental Protection Department, Environmental Restoration Division, Livermore, California, June 11, 1998.

[17] Happel, A., Dooher, B., and Bechenworth, E., Methyl Tertiary-Butyl Ether (MTBE) Impacts to California Groundwater, U.S. EPA Blue Ribbon Panel, Lawrence Livermore National Laboratory, Livermore, California, March 25, 1999.

[18] Notes from California Assembly Natural Resources Committee (Sponsor), Gasoline Oxygenates and New Technology Hearing, Santa Monica, California, November 21, 1997.

[19] Market News Publishing, Methanex Corp Seeks Damages Under NAFTA for California MTBE Ban, www.hydrocarbononline.com, June 16.

[20] Oxygenated Fuels Association, www.ofa.net, 2000.

chapter three

Physical and chemical properties of MTBE

3.1 Production and use of MTBE

Methyl *tertiary*-butyl ether (MTBE) is the most common fuel oxygenate and octane booster currently used in unleaded gasoline in the U.S.[1] In 1997, the annual production of MTBE in the U.S. was about 2.9 billion gallons. The U.S. also imported about 1.2 billion gallons, for a total consumption of 4.1 billion gallons. California's production of MTBE in 1997 was about 181 million gallons. California also imported 922 million gallons of MTBE, for a total consumption of 1500 million (1.5 billion) gallons.[2] In 1997, California produced approximately 5% of U.S. consumption, or about 100,000 barrels per day; 85% of the total amount of MTBE consumed in the U.S. was imported by tankers from overseas producers.[3]

MTBE is manufactured from isobutene (isobutylene or 2-methylpropene: $(CH_3)_2CCH_2$), a byproduct of petroleum refining. Isobutene, a colorless flammable gas with a boiling point of –7 C, is also used in gasoline.[4] MTBE synthesis involves combining isobutene and methanol; MTBE can also be prepared from methanol and *tertiary*-butyl alcohol. Therefore, MTBE can be easily and cheaply produced at the refinery. Because it also easily blends with gasoline, it can be transferred through existing pipelines.[5] Figure 3.1 shows a graphic representation of the MTBE molecule.

The chemical and physical properties of MTBE are summarized in Figure 3.2. The properties of MTBE are described variously by different authors, depending on the precision and accuracy of measuring instruments and variations in laboratory techniques and testing methodology.

More information about the properties of MTBE and gasoline components, as well as guidelines for safe handling, can be found in Appendix C, Material Safety Data Sheets: MTBE and Gasoline, of this publication.

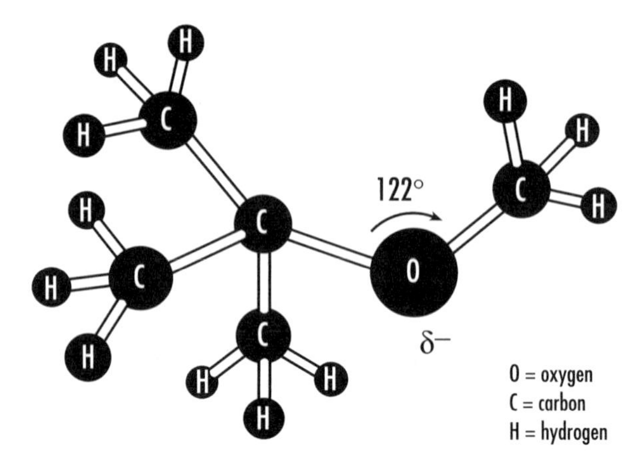

Figure 3.1 Molecular and structural formula of MTBE.

3.1.1 Chemical and structural formula

MTBE is an ether with a general chemical formula of $C_5H_{12}O$[6] and a structural formula of $CH_3OC(CH_3)_3$ or $(CH_3)_3COCH_3$[5,7] as shown in Figure 3-1. The horizontal CH_3-O-C bond represents the ether molecule, and the vertical CH_3-C-CH_3 is a propane molecule; the carbon-oxygen-carbon ether molecule is polar. Therefore, MTBE is sometimes referred to as 2-methoxy 2-methyl propane.[9] At standard temperature (25 C) and pressure (760 mmHg), MTBE is a colorless, flammable, and combustible liquid. Because of its oxygen content, up to 15% (by volume) of MTBE can be added to gasoline to reduce carbon monoxide emissions in internal combustion engines.[3]

3.1.2 Molecular mass

The molecular mass of MTBE is 88.15 g/mole. In general, hydrocarbon compounds with molecular mass less than 150 g/mole can be expected to be quite volatile and will have low melting and boiling temperatures, high vapor pressures, and low adsorption coefficients. (The adsorption coefficient is a chemical's ability to sorb, or bind, to soil particles — the greater the coefficient, the greater the chemical's ability to sorb to soil.) These compounds will readily migrate (volatilize) to the atmosphere from a liquid state.

3.1.3 Melting and boiling points of MTBE

MTBE is a liquid that melts at –109 C and boils at between 53.6 and 55.2 C at standard atmospheric pressure. The melting and boiling points of a substance provide an indication of the physical state of the chemical (e.g., solid,

CAS Registry Number	1634-04-4			
Molecular Formula	$C_5H_{12}O$			
Synonyms	tertiary-butyl methyl ether, t-butyl methyl ether, methyl t-butyl ether, 2-methyoxy 2 methyl propane, 2-methyl-2 methoxypropane, methyl-1, 1-dimethylethyl ether, MTBE			
Beilstein Reference	1,381			
Chemical & Physical Properties	Author			
	Howard (1990)	Squillance et al (1995)	Dean (1992)	Zogorski et al (1996)
Boiling Point (°C)	55.2	—	56	53.6-55.2
Melting Point (°C)	–109	—	–109	—
Flash Point (°C)	—	—	–10	—
Molecular Mass g/mole	88.15	—	—	88.15
Specific Density of Gravity (water=1)	—	—	0.758	0.744
Water Solubility (mg/L) at 25° C	51,000	23,200 to 54,000	—	43,000 to 54,000
Vapor Pressure (V_P) at 25° C	249 mm Hg	3.27 to 3.35×10^4	—	245 to 256 mm Hg

Figure 3.2 Summary of physical and chemical characteristics of MTBE.

liquid, or gas) under standard atmospheric pressure and temperature. Pure MTBE is a colorless, volatile, and flammable liquid.

3.1.4 Volatilization rate of MTBE from water or soil

When MTBE is released to the atmosphere, it tends to occur almost entirely in the vapor phase. MTBE would be expected to have a relatively short volatilization half-life, or $t_{1/2}$, in surface water.[9] MTBE generally has a $t_{1/2}$ in surface water of approximately 9 hours.[10] This half-life, however, can range from 4 weeks to 6 months,[1] depending on the type of surface water. MTBE's $t_{1/2}$ in streams and rivers is lower, because the water is in turbulent flow. MTBE in lake and reservoir water will have a higher $t_{1/2}$ because the water is not being agitated. More MTBE will volatilize from a rapidly flowing river or stream than from relatively quiescent, or slow-moving, lake and reservoir water.

If spilled to surface soil, MTBE volatilizes readily. Additionally, MTBE has a high mobility downward in soil because it does not readily adsorb, or attach, on soil. Therefore, MTBE can reach groundwater relatively quickly and easily.[1]

3.1.5 Solubility

MTBE is readily soluble, or dissolved, in other substances such as alcohol, ether, and gasoline.[7] More importantly, MTBE is highly soluble in water at standard temperature and pressure.[5]

3.1.6 Specific density of MTBE

The specific density of a liquid is represented by the ratio of the density of the substance to the density of water. (The density of water is represented as 1.0.) MTBE has measured specific densities ranging between 0.744[11] and 0.758.[7]

The specific density of a substance determines whether the nondissolved portion of a substance will sink or float if spilled to the water surface. Substances with specific densities greater than 1.0 will sink, and substances with specific densities less than 1.0 will tend to stay above the water's surface ("float"). Once MTBE saturation in water occurs, excess MTBE, like other nondissolved components of gasoline saturated in water, will float on the water's surface.

3.1.7 Henry's law constant for MTBE

Partitioning of a contaminant between the liquid phase and the gaseous phase is governed by Henry's law. Henry's law determines the tendency of a contaminant to volatilize from groundwater into the soil gas. The ratio of the partial pressure of a chemical compound in the vapor phase to the concentration of the compound in the liquid phase at a specific temperature is known as the Henry's law constant of the compound. This constant is a parameter that reflects the air-to-water partitioning of the compound, and is therefore more appropriate than either vapor pressure or water solubility alone for estimating the tendency for volatilization from water to air. Such information is helpful in understanding the phase (water or vapor) in which an organic compound would most likely be found and the relative concentrations of the compound in water or vapor. The Henry's law constant of a substance is often expressed as the ratio of the saturated vapor pressure to the water solubility.

Squillace et al.[5] indicate that a compound with a Henry's law constant of $5 \cdot 10^{-2}$ or larger is termed to be very volatile from water, whereas a compound with a lower value tends to remain in the water phase or partition strongly from the gas phase to the water phase if contaminated vapor contacts water. Based on this classification scheme, the Henry's law constant of MTBE and other fuel oxygenates indicates these gasoline constituents would partition substantially into water.

3.1.8 Human taste and odor thresholds

A taste and odor threshold for MTBE in drinking water occurs at concentrations between 45 and 95 parts per billion (ppb). These concentrations are less than the likely threshold for human chronic injury.[12] Other research suggests that the distinctive, unpleasant odor of MTBE can be detected by most humans at concentrations as low as 20 ppb, substantially less than concentrations known to cause toxic effects in animals.[13] In one case, users of an MTBE-impacted water supply complained of undesirable taste and odors when MTBE concentrations in water were as low as 5 to 15 ppb.[14]

MTBE smells like a terpene, or turpentine, although some who have sampled MTBE report the taste as medicinal, citrus, or even mint-like.[15]

Endnotes and references

[1] Howard, P.H. et al., *Handbook of Environmental Fate and Exposure Data for Organic Chemicals*, Volume IV - Solvents 2, Lewis Publishers, Boca Raton, Florida, 1993.

[2] California Department of Health Services, Does California Need MTBE? at http://www.en.ca.gov/ftp/gen/sorb/mtbe/environ/98 mtbe, 1999.

[3] California Environmental Protection Agency (Cal EPA), MTBE (Methyl *Tertiary*-Butyl Ether): Cal/EPA Briefing Paper, April 4, 1997 (updated June 2, 1997).

[4] Parker, S.P., Ed., *McGraw-Hill Dictionary of Chemical Terms*, McGraw-Hill, San Francisco, California, 1984.

[5] Squillace, P.J. et al., A Preliminary Assessment of the Occurrence and Possible Sources of MTBE in Groundwater of the United States, 1993-94, U.S. Geological Survey Pen-File Report 95-456, 1996.

[6] Howard, P.H. et al., *Handbook of Environmental Degradation Rates*, Lewis Publishers, Chelsea, Michigan, 1991.

[7] Dean, J.A., *Lange's Handbook of Chemistry*, 14th ed., McGraw Hill, New York, 1992.

[8] Garrett, P. et al., MTBE as a Groundwater Contaminant: Proceedings of the NWWA/API Conference on Petroleum Hydrocarbons and Organic Chemicals in Groundwater — Prevention, Detection and Restoration, Houston, Texas, November 12-14, 1986.

[9] A volatilization half-life of MTBE from soil or water represents the time required for half of the quantity of MTBE originally present in the soil or water to be lost to the atmosphere.

[10] Pankow, J.F. et al., Calculated volatilization rates of fuel oxygenate compounds and other gasoline-related compounds from rivers and streams, *Chemosphere*, 33, 5, 1996.

[11] Zogorski, J.S. et al., Fuel Oxygenates and Water Quality: Current Understanding of Sources, Occurrence in Natural Waters, Environmental Behavior, Fate, and Significance, Interagency Oxygenated Fuel Assessment, Office of Science and Technology, Washington, D.C., 1996.

[12] Tardiff, R.G., Estimating the Risks and Safety of Methyl-*Tertiary*-Butyl Ether (MTBE) and *Tertiary* Butyl Alcohol (TBA) in Tap Water for Exposures of Varying Duration, Division of Environmental Chemistry Preprints of Extended Abstracts, v. 37, n. 1, April 1997.

[13] Stelljes, M., Issues Associated with the Toxicological Data on MTBE, Division of Environmental Chemistry Preprints of Extended Abstracts, v. 37, n. 1, April 1997.

[14] Davidson, J.M., Fate and Transportation of MTBE — The Latest Data, National Water Well Association, Proceedings of the Petroleum Hydrocarbon and Organic Chemicals in Groundwater, Prevention, Detection and Remediation Conference, Nov. 29 – Dec. 1, Houston, Texas, 1995.

[15] Von Burg, R., Toxicology Update: Methyl tert-Butyl Ether, *J. Appl. Toxicol.*, 12, 1, 1992.

chapter four

Toxicity, health effects, and taste and odor thresholds of MTBE

4.1 Introduction

After the introduction of oxygenated gasoline containing MTBE in the United States, acute health system complaints, such as headaches, nausea, dizziness, and breathing difficulties were reported both regionally and nationwide. Existing studies of the acute health risks of MTBE, however, generally don't support the claims that the use of MTBE in gasoline causes significant increases in acute symptoms. There are no studies on MTBE's carcinogenicity in humans, but studies indicate that MTBE is a carcinogen to rats and mice. The U.S. Environmental Protection Agency (U.S. EPA) has classified MTBE as a possible human carcinogen.[1] Although some studies suggest that MTBE is not a human carcinogen, the Office of Environmental Health Hazard Assessment (OEHHA) has set a public health goal of 13 parts per billion (ppb) for MTBE in drinking water based on cancer studies in rats and mice.[2] However, the levels at which MTBE potentially poses a cancer or noncancer human health risk are likely far higher than the levels at which humans can detect the taste of MTBE in drinking water. (See also Chapter 3, Physical and Chemical Properties of MTBE, for more discussion of the taste and odor thresholds of MTBE.)

In 1999, California's Department of Health Services (DHS) proposed a 13 micrograms per liter (mg/L) or 13 ppb primary maximum contaminant level (MCL) for MTBE.[3] A contaminant's primary MCL indicates the concentration that is not to be exceeded in statewide public water supplies. The secondary MCL for MTBE has been set at 5 mg/L. This secondary MCL for MTBE is known as the taste and odor threshold — this threshold also indicates a concentration that is not to be exceeded in statewide public water supplies.

This chapter explores the potential for MTBE to pose a cancer or noncancer human health risk, according to existing but limited data. This chapter also compares the level at which MTBE is detected by taste or odor to the level or concentration of MTBE in drinking water that may pose a threat to human health.

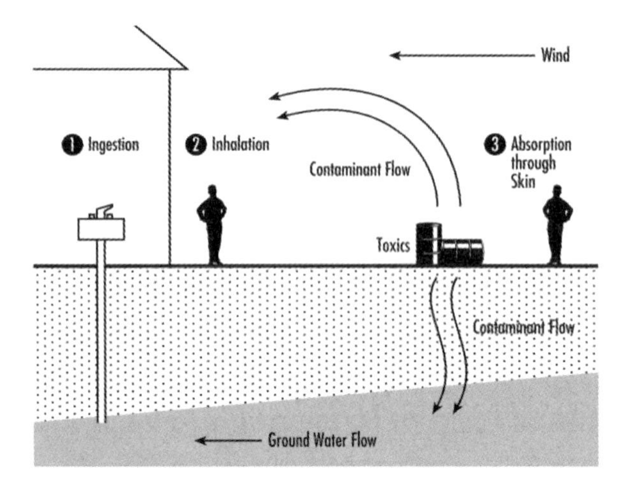

Figure 4.1 Sources and receptors of pollutants.

MTBE does not stay in the body long; it is released through exhalation and urine excretion. Following an exposure to MTBE, most of the substance will leave the body in about 2 days. The MTBE that is not released from the body is transformed (mostly through hydrolysis) into other compounds such as acetone, *tertiary*-butyl alcohol (TBA), methyl alcohol, formaldehyde, and carbon dioxide. Formaldehyde is classified by the U.S. EPA as a probable human carcinogen, and there is some evidence that TBA is an animal carcinogen in male rats and female mice.[4]

4.2 Toxicology

For any substance to have an adverse health effect, it must first enter the body. The common exposure routes or intake modes are

- Ingestion (eating/drinking)
- Dermal contact (skin penetration)
- Inhalation (breathing)

Exposure routes and pathways are conceptualized in Figure 4.1. Health effects can be characterized as carcinogenic (cancer-causing) and noncarcinogenic, and by the following three exposure times resulting in an adverse effect:

- Acute (14 days or less)
- Intermediate (15 to 364 days)
- Chronic (365 days or greater)

From the vantage of assessing MTBE in groundwater, this chapter is concerned generally with ingestion as the primary mode of human exposure through drinking contaminated water or eating affected animals.

For more specific information regarding calculations for the human health risk of MTBE, refer to Appendix I, Toxicity of MTBE: Human Health Risk Calculations, of this publication.

4.2.1 Ingestion

Ingestion of MTBE can occur either acutely or chronically, but an evaluation of the possibility for long-term ingestion of contaminated drinking water is of the greatest relevance to the general public. All public drinking water providers perform testing on their water sources. The longest-term human ingestion of water potentially contaminated with MTBE would not be any greater than 3 to 6 years (the public water supply sampling cycle period for chemicals considered low priority in terms of human health risk; MTBE has tended to be classified in this category).

4.2.2 Dermal contact

Dermal contact is a secondary mode of human exposure and can occur through bathing or washing in contaminated waters. Dermal contact with fuel additives probably occurs as an infrequent acute exposure for the general population, with the exception of auto mechanics and service station attendants.

4.2.3 Inhalation

Inhalation, although an important mode of exposure, tends to be infrequent, and most likely associated with vehicle refueling, either by self-service gasoline customers or service station attendants. Various studies of personal breathing zone samples of MTBE during gasoline refueling suggest that such airborne exposures typically amount to 2 to 5 minutes in duration, and may range as high as 2 to 32 parts per million by volume (ppmv) MTBE. Most of the data for exposure during refueling suggest that these airborne exposures tend to be less than 10 ppmv during the 1- to 20-minute sampling periods.[5] Inhalation of airborne MTBE associated with MTBE-contaminated soil or groundwater is an unlikely source of exposure for the general public, and unlikely to be associated with soil and groundwater issues.

Exposure to airborne MTBE through inhalation such as while showering may occur as well, in the context of exposure to tap water. (Elevated water temperatures used during bathing and showing allow for higher volatilization rates of MTBE and consequently a highere inhalation potential.)

4.3 Cancer effects

Of the cancer studies conducted on MTBE, the substance was shown to cause cancer in rats through inhalation exposure at the two highest tested dose levels (10,800 mg/m^3 and 28,800 mg/m^3, respectively).[6] Similarly, inhalation of MTBE caused cancer in mice at the highest dose level in a second study.[7] Based on the authors' reviews of the available literature, including a review of the study conducted by the OEHHA,[2] no rigorous, peer-reviewed epidemiological studies (or any other studies) exist indicating a human cancer risk from exposure to MTBE. Extrapolation of data on animal toxicity from inhalation, to data on human health risk from ingestion or dermal contact, is ambiguous at best; and the validity of any conclusions drawn is reduced by inherent uncertainties. Moreover, the mechanism for MTBE's toxicity in rats and mice does not appear to be relevant in humans.[9] Nevertheless, the U.S. EPA has classified MTBE as a Group C chemical, or possible human carcinogen,[1] and OEHHA has established a public health goal of 13 ppb for MTBE in drinking water, as discussed previously in this chapter.

4.3.1 Inhalation

No information exists regarding cancer in humans from inhalation of MTBE, because no studies have been conducted on MTBE's carcinogenicity in humans.

Renal (kidney) tumors and lesions have occurred in male rats from inhalation exposure to concentrations of 3000 and 8000 ppmv MTBE. However, this effect may be caused by MTBE's stimulation of the accumulation of one or more proteins unique to the male rat. Hepatocyte (liver) tumors have occurred in female mice from inhalation exposure to concentrations of 8000 ppmv MTBE. As with the rat, this effect in female mice may be unique to female mice. Based on this animal data, and until more data become available, the U.S. EPA cautiously considers MTBE a weak inhalation carcinogen to humans, and has provided a preliminary cancer unit risk for MTBE. By comparison, benzene, a component of gasoline, has a U.S. EPA-designated inhalation cancer risk number about 55 times greater than MTBE.

4.3.2 Ingestion

No information exists regarding cancer in humans from ingestion of MTBE, because no studies have been conducted on MTBE's carcinogenicity in humans.

Leukemia occurred in female rats to which MTBE was administered at a dose of 1000 mg per kg body mass per day (mg/kg body mass/d) for 104 days. Testicular tumors were observed in male rats given a dose of 1000 mg/kg body mass/d MTBE for 104 days.

4.3.3 Dermal

No information exists regarding cancer in humans or animals from dermal exposure to MTBE.

4.4 Noncancer effects

4.4.1 Inhalation

No deaths in humans have been reported due to inhalation of MTBE. No deaths occurred in rats or mice exposed to levels of MTBE greater than 8000 ppmv for 6 hours per day, 5 days per week, for 4 or 5 weeks.

4.4.1.1 Systemic–immunologic/lymphoretic–neurologic

Although several epidemiological studies have reported adverse systemic effects (such as dizziness, headache, and nausea) from inhalation exposure to ambient concentrations of MTBE, it has not been possible to statistically correlate exposure to ambient or occupational concentrations of MTBE with the alleged acute effects on humans. Two studies which reported adverse systemic effects to humans through inhalation of MTBE indicated 1.5 ppmv MTBE as the lowest level at which inhalation resulted in no-observed-adverse-effects in humans.

4.4.2 Ingestion

No deaths in humans have been reported from ingestion of MTBE. The ingestion (by gavage) to a lethal dose for 50% of the tested population (LD_{50}) for the rat and mouse is about 4000 mg/kg body mass/d. No deaths have occurred in rats and mice from ingestion of MTBE at maximum doses of 1000 mg/kg body mass/d. This indicates that MTBE has a low ingestion toxicity in experimental animals.

4.4.2.1 Systemic–immunologic/lymphoretic–neurologic

No information exists regarding systemic effects in humans from ingestion of MTBE. A lowest ingestion No-Observed-Adverse-Effect-Level (NOAEL) of about 70 mg/kg body mass/d and a lowest ingestion Lowest Observed Adverse Effect Level (LOAEL) of about 100 mg/kg body mass/d have been observed for the rat.

Some research shows that humans can begin to smell MTBE in water at a concentration of less than 20 ppb, and to taste MTBE in water at a concentration of about 40 ppb under ambient conditions. (Refer also to Chapter 3, Physical and Chemical Properties of MTBE, of this publication for more detail regarding taste and odor thresholds for MTBE.) Because the taste and odor of MTBE can be detected at these low levels, such detection can provide a "warning" that can prevent accidental ingestion or inhalation of health-threatening concentrations of MTBE in water.

4.4.3 Dermal

No deaths in humans have been reported from dermal exposure to MTBE. In animal studies, no deaths occurred in rats from a 6-hour dermal exposure to MTBE at a concentration of 400 mg/kg body mass/d.

4.4.3.1 Systemic–immunologic/lymphoretic–neurologic

In animal studies, no systemic adverse effects were produced after direct application of liquid MTBE to the skin. Exposure of skin to airborne MTBE does not produce any effects to humans or laboratory animals.

4.5 Ecological effects: aquatic toxicity

The first aquatic toxicity data for MTBE were published in the early 1980s for representative fresh water and marine species.[9,10] Bioconcentration studies with Japanese carp generated bioconcentration factors for MTBE ranging from 0.8 to 1.5, and concluded that MTBE does not bioconcentrate appreciably.[11] Currently, no national ambient water quality criteria or state water quality standards for the protection of aquatic species exist for MTBE.[12] Future regulatory efforts, such as through the National Pollution Discharge Elimination System (NPDES) discharge and storm water discharge monitoring process, will likely be used to assess levels of MTBE in aquatic ecosystems.

4.6 Evaluation of studies and data gaps

The current data appear adequate for evaluating the toxicity of MTBE to experimental rats and mice. That is not true in the case of human exposure to MTBE. While some toxicological information exists for noncarcinogenic health effects to humans from inhalation of MTBE, little to no information exists regarding reproductive, developmental, genotoxic, and carcinogenic effects. Also, no assessment of the quality or accuracy of the published toxicological data currently exists.

Clearly, there is a need to determine the relevance of MTBE's carcinogenic effects observed in rats and mice to its potential as a carcinogen in humans. Based on the toxicological mechanism involved in rats and mice, it is possible and even likely that the observed carcinogenic effects are not relevant to any human cancer risk. Studies on the systemic adverse human health effects of MTBE, such as reproductive, developmental, and genotoxic effects, are also needed.

And, because most human exposure to MTBE occurs from the inhalation of gasoline vapors, there is a need to compare the toxicity of the gasoline plus MTBE additive mixture to that of gasoline without MTBE.

Moreover, any estimates of cancer risk in humans from exposure to MTBE based on or derived from animal studies are subject to considerable uncertainty.

4.7 Key issues

Although more studies are needed to determine the cancer and noncancer human health risks of MTBE, most existing relevant literature suggests that MTBE may not be as toxic as anecdotal stories of acute health effects from MTBE exposure seem to suggest. Health data suggest that MTBE has at least 55 times less carcinogenic toxicity than benzene, another component of gasoline. Moreover, it is likely that MTBE's unpleasant taste and odor, detectable at very low levels, would provide a "warning" to prevent accidental ingestion of health-threatening concentrations of MTBE in water.

Endnotes and references

Unless otherwise noted, the information in this section on toxicity is based on 1) The Toxicological Profile for Methyl tert-Butyl Ether, U.S. Department of Health & Human Services, Public Health Service, Agency for Toxic Substances and Disease Registry (ATSDR), 1996; and 2) United States Environmental Protection Agency (U.S. EPA), Integrated risk information system, On-Line Database, January 1996.

[1] U.S. EPA, 1996.

[2] Office of Environmental Health Hazard Assessment (OEHHA), Public Health Goal for Methyl Tertiary-Butyl Ether (MTBE) in Drinking Water, March 1999. OEHHA, 1999.

[3] mg/L are generally equivalent to ppb (in water)—see also Appendix B, Conversions Table, of this publication.

[4] Zogorski, J.S. et al., Fuel Oxygenates and Water Quality: Current Understanding of Sources, Occurrence in Natural Water, Environmental Behavior, Fate and Significance, Interagency Oxygenated Fuel Assessment, Office of Science and Technology, Washington, D.C., 1996.

[5] Melnick, R.L. et. al., Potential Health Effects of Oxygenated Gasoline, National Science and Technology Council, Interagency Assessment of Oxygenated Fuels, Committee on Environmental and Natural Resources, Executive Office of the President of the United States, Washington, D.C., June 1997.

[6] Chun, J.S. et al., Methyl tertiary-butyl ether; vapor inhalation oncogenicity study in Fischer 344 rats, Bushy Run Center, Export, Pennsylvania, Project No. 91N0013B, 1992.

[7] Bureigh-Flayer, H.D. et al., Methyl tertiary-butyl ether; vapor inhalation oncogenicity study in CD-10 mice, Bushy Run Center, Export, Pennsylvania, Project No. 91N0013A, 1992.

[8] See the toxicological profile provided by the Agency for Toxic Substances and Disease Registry (ASTDR), 1996.

[9] Veith, G.D. et al., Estimating the acute toxicity of narcotic industrial chemicals to fathead minnows, in *Aquatic Toxicity and Hazard Assessment*, Bishop, W.E., R.D. Cardwell, and B.B. Heidolph, Eds., Philadelphia, Pennsylvania, 6th Symposium, ASTM STP 802, p. 90, 1983.

[10] Bengsston, E.B. and M. Tarkpea, The acute aquatic toxicity of some substances carried by ships, *Marine Pollution Bulletin*, v. 14 n. 6, 1983.

[11] Fujiwara, Y. et. al., Biodegradation and bioconcentration of alkyl ethers, *Yukayaken*, 33, 1984.

[12] Stubblefield, W.A. et al., Evaluation of The Acute and Chronic Aquatic Toxicity of Methyl Tertiary-Butyl Ether (MTBE), Division of Environmental Chemistry Preprints of Extended Abstracts, v. 37 n. 1, April 1997.

chapter five

Transport and fate of MTBE in the environment

5.1 Introduction

The main contributors of gasoline containing MTBE into the subsurface environment of soil and groundwater are underground storage tanks, pipelines, and refueling facilities. Large quantities of MTBE may potentially be released from these sources. In some cases, such as in Santa Monica, California, the detection of high concentrations of MTBE in drinking water sources (determined to be from several leaking underground storage tanks) has resulted in shutdown of water supply wells and high cleanup costs. Figure 5.1 illustrates the release of gasoline containing MTBE into the groundwater and public water supply.

Natural attenuation is increasingly being used as a remediation strategy for sites contaminated with fuel hydrocarbons. Natural attenuation allows natural chemical processes to degrade contaminants, as opposed to the active removal of contaminants through engineered solutions. The presence of MTBE at a gasoline-contaminated site, however, complicates this approach because MTBE resists natural biological degradation and can inhibit degradation of other fuel components. Additionally, MTBE moves more quickly through groundwater than other gasoline fuel constituents, such as benzene. The rapid movement of MTBE in groundwater further complicates natural attenuation strategies for management of MTBE-contaminated sites.

Efficient and cost-effective assessment and remediation of sites contaminated by subsurface petroleum hydrocarbon constituents require a fundamental understanding of both the source of the contaminants and their transport and fate in the subsurface. The purpose of this chapter is to present a review of the existing knowledge of MTBE transport and fate in the environment, so that the information may be used to inform examinations of environmental contamination management.

Information on other oxygenates — including ethanol, ethyl *tertiary*-butyl ether (ETBE), *tertiary*-amyl methyl ether (TAME), di-isopropyl ether (DIPE), and their degradation products — is also reviewed in this chapter. Methanol and *tertiary*-butyl alcohol (TBA) are also addressed because they

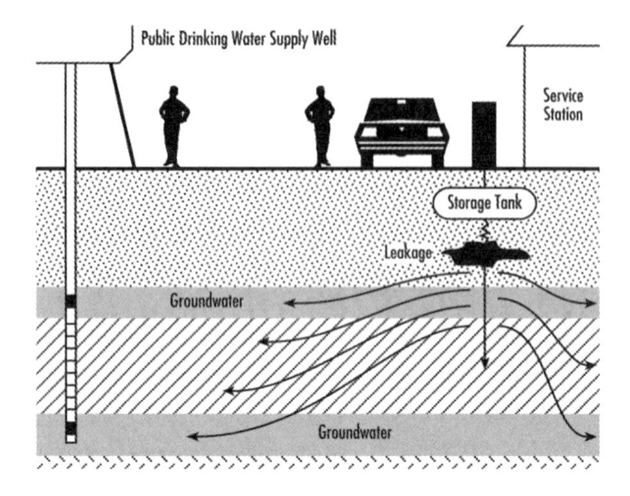

Figure 5.1 How MTBE gets into groundwater.

have been used as gasoline additives in the past, and may be used in the future. Other more standard gasoline components such as benzene are also discussed for purposes of comparison.

5.2 *Importance of transport and fate assessments in environmental restoration projects*

To achieve environmental restoration, knowledge of the transport and fate of contaminants in the subsurface environment is essential. Without this knowledge, inadequate cleanup of contamination, or costly and unnecessary preventative actions and responses to contamination, can result. However, gaining and using knowledge about transport and fate can be difficult because of the complexities of site hydrogeology, hydrochemistry, and microbiology.

The concentrations and migration rates of hazardous substances in groundwater play major roles in determining how far hazardous substances migrate, how to assess risk factors, and what remedial action should be taken at a specific site. These analyses require a thorough understanding of the geochemical behavior of these substances in soil-water systems.

Chemicals, especially organic compounds, are subject to a variety of processes in the subsurface that affect their transport and fate to a certain degree, depending upon site conditions and the specific chemical compounds in question. The major parameters affecting chemical transport in groundwater are adsorption and biodegradation. These parameters will be discussed further in this chapter.

For an in-depth discussion of the geological aspects of the subsurface environment, refer to Appendix E, Geologic Principles and MTBE, of this publication. For more discussion and graphic illustration of the transport and fate of MTBE in the subsurface environment, refer to Appendix H, Plume Geometries for Subsurface Concentrations of MTBE, of this publication.

5.2.1 *Transport and fate assessments in environmental restoration*

The three main phases of environmental restoration programs are site characterization, risk assessment, and remediation. Many conventional approaches to site characterization do not adequately emphasize the need to obtain detailed information about natural processes affecting the transport behavior and the ultimate fate of contaminants. The use of state-of-the-art site characterization tools such as cone-penetrometer testing (CPT), logging by CPT or direct push (Geoprobe®) rigs and rapid petroleum hydrocarbon delineation by the rapid optical screening tool (ROST®) may be worthwhile, although they are more costly to implement than more conventional direct push (Geoprobe®) or hollow stem auger sampling using a rig. These newer site assessment and characterization tools can ultimately result in significant savings because of their improved technical effectiveness and the efficiency of their use in site cleanup for larger sites. Proper site characterization methods can aid risk assessments and risk management decisions in determining if remediation is necessary, and in choosing appropriate cleanup technologies.

For a discussion of the phases of environmental restoration programs and of various protocols for site assessment, please refer to Appendix F, MTBE: Subsurface Investigation and Cleanup, of this publication.

5.2.1.1 *Site characterization*

The use of a sound subsurface conceptual model is necessary for site characterization. Such a model must incorporate information on geological, hydrological, chemical, and biological processes effectively into the transport and fate evaluation of a contaminant. Such information requires data that are accurate and appropriate for the intended assessment objectives. The process of data collection and evaluation for site characterization is made difficult by the potential multitude of chemicals involved and the complex hydrogeological setting of a site.

Another of the main difficulties in using site characterization data for designing a remediation plan is with scale — i.e., moving from a large scale of site characterization to a much smaller scale of a specific site where remediation is required.

5.2.1.2 *Risk assessment and remediation*

A major issue in groundwater remediation is assessing the risks of hazardous chemical exposure to the public and the environment, within the limitations of remedial technologies. This issue is considered in determining the concentrations to which contaminants are remediated. In remedial actions, contamination concentrations can be dramatically reduced after a reasonable period of time. However, the contaminant mass reduction rate usually declines with time, and gradually approaches a residual (or asymptotic) concentration, which may or may not be less than the cleanup levels established for the site based on available remedial technologies and environmental risk assessment.

The persistence of residual contamination can be explained through an understanding of the numerous physicochemical parameters involving subsurface hydrogeology and the types of contaminants at the site. For example, rapid cleanup of dissolved contaminants by groundwater extraction and treatment can be accomplished in zones of the highest hydraulic conductivities; however, cleanup in zones with lower hydraulic conductivities can sometimes only occur with diffusion of the contamination out of the low conductivity zones. Also, many contaminants readily adsorb (attach) to aquifer materials because of the nature of the soil/rock material and the properties of the contaminants. This is a significant problem at sites with heterogeneous hydrogeologic settings.

5.2.2 Transport and fate processes

The environment comprises a number of different compartments including water, soil, and air. The ways in which MTBE and other fuel compounds behave in the environment are determined by how they distribute themselves among the different possible compartments. The following important components of contaminant transport and fate processes are addressed in this chapter:

- Physical and chemical (physicochemical) characteristics of contaminants.
- Physical processes controlling dissolved contaminant transport and fate.
- Chemical processes controlling dissolved contaminant transport and fate.
- Biodegradation.

5.3 Physical and chemical characteristics of MTBE important to its transport and fate

For a more in-depth description of the physicochemical characteristics of MTBE, refer to Chapter 3, Physical and Chemical Properties of MTBE; as well as Appendix G, Synthesis, Properties, and Environmental Fate of MTBE and Oxygenate Chemicals, of this publication.

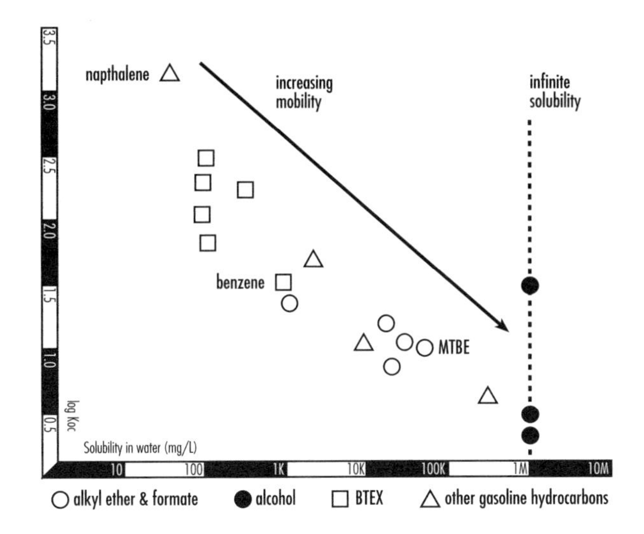

Figure 5.2 Mobility of gasoline hydrocarbons in groundwater.

5.3.1 Vapor pressure

MTBE has a relatively high vapor saturated pressure and low boiling point. In soil, MTBE exists only to a small degree as a soil gas.

5.3.2 Water solubility

MTBE has a low volatility in water. This volatility is depressed because MTBE and water form hydrogen bonds. This hydrogen bonding also results in MTBE's high solubility in water. Highly soluble substances can be leached rapidly from soils and are generally mobile in groundwater. Figure 5.2 shows a comparison of the mobility of various gasoline hydrocarbons in groundwater.

5.3.3 Henry's law constant

Whether a contaminant exists predominantly as a liquid or a gas is governed by Henry's law, the application of which can predict the tendency of a contaminant to volatilize from groundwater into the soil gas.*

*Soil gas refers to the vapor phase in the unsaturated, or vadose, zone void space.

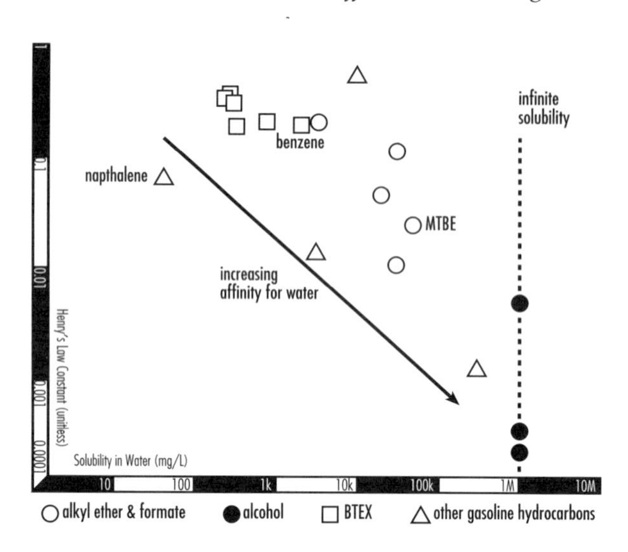

Figure 5.3 Partitioning of gasoline hydrocarbons between water and air.

Henry's law can more successfully estimate the tendency of a contaminant to volatilize from water to air than either saturated vapor pressure or water solubility alone. Such information is helpful in understanding the physical phase (water or vapor) in which an organic compound will most likely be found. Figure 5.3 shows the partitioning of gasoline hydrocarbons between water and air.

For more on the application of Henry's law to MTBE, refer to Chapter 3, Physical and Chemical Properties of MTBE.

5.3.4 *Specific density*

The specific density of a contaminant is extremely important in determining its migration relative to the location of the water-bearing aquifer. The specific density of a substance determines whether the substance will sink or float if spilled to the water surface. Substances with specific densities greater than 1.0 will sink, and substances with specific densities less than 1.0 will tend to stay above the water's surface ("float").

Gasoline as a mixture has a specific density of 0.740; MTBE and the other fuel oxygenates generally have specific densities close to this value. The specific densities of the BTEX (benzene, toluene, ethylbenzene, and xylenes) compounds are generally between 0.86 and 0.88. MTBE that is not dissolved in water, like other nondissolved components of gasoline, will tend to float on the water's surface.

5.3.5 *Cosolvency effects*

A cosolvency effect of a compound describes the potential for the presence of that compound to enhance the subsurface transport velocities of other compounds (e.g., other gasoline constituents). Although concentrations of MTBE can be high in groundwater, they are typically not high enough to significantly increase either the water solubilities or the transport rates of other potentially hazardous gasoline constituent compounds present in groundwater.[1]

5.4 *Transport and fate of MTBE in the atmosphere*

The occurrence of MTBE in the atmosphere is primarily as a result of evaporative emissions and incomplete combustion.[1] Median concentrations of MTBE in urban air have been reported at 3.60 micrograms per cubic meter (mg/m^3) or less.[2] Transport of MTBE in the atmosphere is affected by wind speed, air temperature, and the compound's inherent volatility. MTBE is generally removed from the atmosphere by chemical degradation or precipitation.[1]

5.4.1 *Fate of atmospheric MTBE*

A limited amount of work has been conducted to investigate the atmospheric lifetimes of selected fuel oxygenates.[2]

MTBE, along with other fuel oxygenates, tends to partition into atmospheric water, becoming incorporated into precipitation.[1] Henry's law can be applied to predict this partitioning. Atmospheric precipitation will not result in any significant decrease in the mass of MTBE and other fuel oxygenates in the environment.[2] But such precipitation can transport MTBE to the ground surface and subsurface. This can result in the contamination of surface water and groundwater (see subsection 5.6, Transport and Fate of MTBE in Groundwater, of this chapter).

The major process that governs the fate of fuel oxygenates in the atmosphere is reaction with hydroxyl radicals (OH·).[3] Predicted atmospheric lifetimes of MTBE and TBA in reaction with a given concentration of OH· have been reported as 4 and 11 days, respectively.[4] These atmospheric lifetimes are much shorter than those for MTBE and TBA subject only to the effect of partitioning from precipitation. Reactivity of DIPE with OH· radicals is higher than that for MTBE, ETBE, methanol, or TBA.[3] Isopropyl acetate and formaldehyde are the main byproducts of the reaction, which could become problematic if DIPE is widely used as a fuel oxygenate in the future.

Estimates of the atmospheric half-life of MTBE (the half-life is the lifetime multiplied by the natural logarithm of 2) are as short as 3 days in a regional airshed; however, MTBE could resist degradation in areas of relatively low concentrations of OH· (e.g., metropolitan areas).[2]

5.4.2 *Degradation*

The processes that degrade MTBE in the atmosphere also result in the formation of other chemicals (degradation products). *Tertiary*-butyl formate (TBF) is the most common chemical produced, followed by methyl acetate (acetic acid), acetone, TBA, and formaldehyde.[5,6]

5.5 Transport and fate of MTBE in surface waters

MTBE has been detected in surface waters, including urban storm waters, throughout the U.S. A wide variety of conditions affect the presence and concentration of MTBE and other fuel oxygenates in surface waters, including various atmospheric conditions, gasoline usage, and the surface water's physical characteristics (temperature, velocity, etc.). Three factors primarily affect the transport process of fuel oxygenates in surface waters: water velocity, depth, and temperature. Also, the physicochemical characteristics of the oxygenates play an important role in their transport, especially those governed by the Henry's law constant.

Surface water concentrations of MTBE in California have been documented as high as 12 milligrams per liter (mg/L) in water bodies where small engines from motorized water recreational vehicles are operated. MTBE detected in urban stormwater runoff samples in California has generally been less than 2 mg/L. Higher concentrations have been noted in heavily urbanized areas such as New York City.[7]

5.5.1 Fate of MTBE in surface waters

Under certain conditions, the rate of a chemical's volatilization from surface waters can be independent of the Henry's law constant; chemicals with different Henry's law constants can volatilize to the atmosphere at essentially the same rate under certain flow conditions.[2] Volatilization of fuel oxygenates from flowing surface waters can be controlled by the rate at which the dissolved constituent within the surface water body moves to the water surface. For example, MTBE entering a deep, slow-moving surface water body can remain in the water for several months before volatilizing to the atmosphere.

Squillace et al.[1] presented half-life values of MTBE representing volatilization from a flowing stream or river from studies conducted by Schwarzenbach et al.[8] and Pankow et al.[9] The half-life values are a function of water velocity, water depth, and atmospheric temperature, with no account made for biodegradation. These data are graphically presented in Figures 5.4 and 5.5 for scenarios of winter air and water temperature (Figure 5.4) and summer air and water temperature (Figure 5.5). For both scenarios, the half-life of MTBE increases with increasing water depth and decreases with increasing water velocity. Also, for a given water depth and velocity, the half-life of MTBE is longer under the winter air-temperature scenario than for the summer scenario. The increase in temperature from the winter to summer scenario generally results in a reduction of the half-life (by a factor of 2 to 3). Somewhat greater reductions (shorter half-life scenarios) can be expected in fast-moving, shallow streams and urban runoff channels.

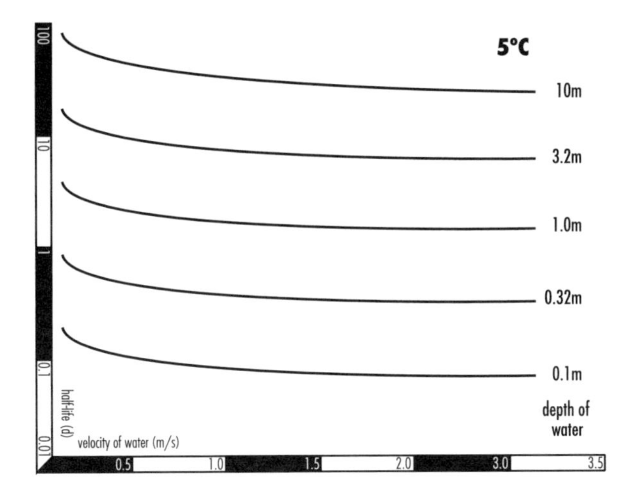

Figure 5.4 Estimated half-life of MTBE for different water depths in a stream/river for calm air at 5°C.

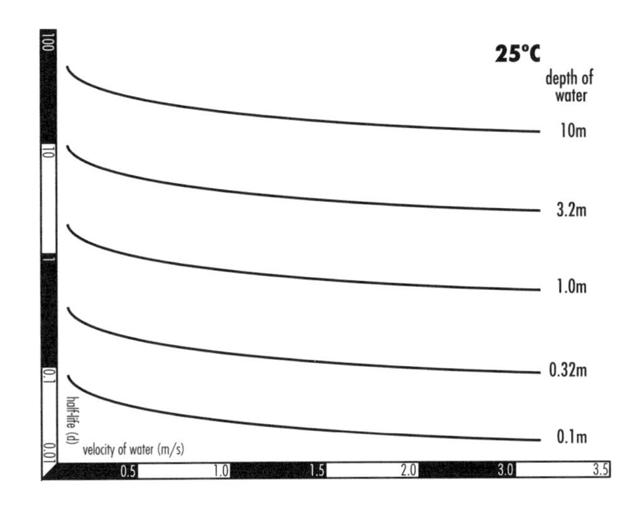

Figure 5.5 Estimated half-life of MTBE for different water depths in a stream/river for calm air at 25°C.

The alkyl ether oxygenates ETBE, TAME, and DIPE can be expected to behave very much like MTBE as described above. TBA will have a significantly higher half-life because of its low Henry's law constant. And even though benzene has a lower Henry's law constant than MTBE, benzene more readily volatilizes from a shallow stream owing to its higher volatility. The volatilization rate increases with an increase in the shallowness of the stream and in the velocity of water movement.

5.6 Transport and fate of MTBE in groundwater

The source of the majority of MTBE in groundwater is from leaking gasoline storage tanks and their associated piping and refueling facilities. Urban and industrial runoff and wastewater discharges are other potential sources of MTBE in groundwater. Atmospheric precipitation in urban areas may be responsible, to a lesser extent, for delivery of MTBE to groundwater as discussed above (see section 5.4, Transport and Fate of MTBE in the Atmosphere, of this chapter). Nonetheless, the processes of infiltration and diffusion of MTBE through the vadose zone into the saturated zone (below groundwater table) are the important transport mechanisms leading to contaminated groundwater. Transport modeling by Pankow et al.[10] showed infiltration, the percolation of water through soil, as the primary mechanism for the transport of MTBE to groundwater. MTBE and other fuel oxygenates can be transported to groundwater much more rapidly via manmade pathways, such as dry wells that recharge storm water directly to groundwater.[2] And relatively high concentrations of MTBE in groundwater can be expected in regions where atmospheric concentrations of MTBE are higher, such as areas with a high density of automobiles; precipitation falling in these areas contains more atmospheric MTBE than in other areas.

MTBE dissolved in groundwater will behave according to its rate of partitioning from gasoline and/or the atmosphere, its solubility, and its interactions with the soil matrix. Based on experimental water-solubility data, high concentrations of MTBE in groundwater are more and more common, especially at sites of a gasoline release.[1] MTBE concentrations as high as 200 mg/L in groundwater have been reported;[11,12] this concentration is far less than that predicted by the effective solubility of MTBE (5000 mg/L). This discrepancy is probably the result of dilution effects or by depletion of MTBE in the gasoline itself by dissolution and/or partitioning.[2]

5.6.1 Adsorption potential (retardation by adsorption)

The slowing of the migration of an organic contaminant in groundwater in relation to the movement of the groundwater itself (also known as retardation) can be caused by many mechanisms, such as adsorption (attachment) of the contaminant to soil particles. Organic compounds like MTBE and BTEX will adsorb to some extent on subsurface solids, especially when such solids have a relatively high content of organic carbon, even though they are highly soluble in water compared to other organic compounds in gasoline. However, MTBE will adsorb much less readily to subsurface solids than will BTEX. Under one theoretical set of conditions, it is estimated that about 8%

of the total MTBE present in an aquifer would be adsorbed to the subsurface material, whereas about 40% of the total benzene would be adsorbed.[1]

5.6.2 Potential for degradation by biotic and abiotic chemical reactions

Degradation of a contaminant in groundwater refers to the natural tendency of the contaminant to be reduced in concentration over time, as a result of subsurface chemical processes. These chemical processes involve either the activity of living organisms such as bacteria (this process is referred to as "biodegradation") or nonbiological chemical processes ("abiotic degradation"). As previously discussed, MTBE dissolved in water does not tend to significantly adsorb to sediments in surface water, nor to the soil matrix in groundwater. Additionally, MTBE does not tend to undergo much degradation due to abiotic reactions such as hydrolysis, photolysis, or photooxidation through reactions with hydroxyl radicals in water.[13] Biodegradation of MTBE is the remaining potential mechanism of retardation, and is explored in the following sections of this chapter.

5.6.3 Potential for biodegradation

The potential for biodegradation of fuel oxygenates, including MTBE, has been the subject of only a limited amount of research. Results of studies to date generally show that MTBE and other alkyl ether oxygenates (TAME, ETBE, DIPE) biodegrade with difficulty, whereas ethanol, methanol, and BTEX biodegrade readily.[2] MTBE has been classified by some authors as recalcitrant,[14–17] owning perhaps to the typical behavior of the carbon atomic structure and/or the ether structure.[18] For further information regarding the transport and fate implications of the molecular structure of MTBE, refer to Appendix G, Synthesis, Properties, and Environmental Fate of MTBE and Oxygenates, in this publication.

Some research has indicated a low rate or lack of MTBE biodegradation under anaerobic (without oxygen) conditions in microcosms commonly found in landfill material, soils, and sludges.[19–22] Also, a study by Yeh and Novak[37] reported no observed degradation of MTBE in aerobic microcosms.

One study, however, reported biodegradation of MTBE in the source area of a fuel spill[23] based on field evidence and laboratory studies. More examples of MTBE biodegradation have been reported in studies of oxygen-limited microcosms.[16,24] TBA was observed as a persistent degradation product in these studies. Other laboratory studies[21] have reported biodegradation of MTBE under anaerobic conditions, but only under specific conditions of pH, coupled with a lack of specific organic matter type.[1]

5.6.4 Biodegradation of other fuel oxygenates

ETBE, while also fairly recalcitrant, has been reported as being less resistant to biodegradation than MTBE.[2] Anaerobic biodegradation of ETBE was observed in one study after a lag period; the degradation rate decreased owing to an accumulation of TBA and an increase in pH.[21] Other studies have reported relatively slow rates of anaerobic biodegradation of ETBE with no production of TBA under limited conditions, and rapid aerobic biodegradation of ETBE with TBA accumulation.[17]

TAME's lack of biodegradability has been shown under various conditions,[16] including studies of TAME in the presence of aquifer material, soil, or activated sludges.[20] Similar results have been reported in studies of the susceptibility of DIPE to anaerobic biodegradation.[15,16,19]

As previously mentioned, TBA can be a byproduct of the biodegradation of ETBE. TBA has been reported to be resistant to biodegradation.[15,25] However, anaerobic biodegradation of TBA has been reported in certain conditions depending on the TBA concentration and indigenous microbial activity.[2]

TBF, one of the main byproducts of MTBE degradation in the atmosphere, is a potential groundwater contaminant. After its generation in the atmosphere through the process of photooxidation (along with acetone, formaldehyde, methyl acetate, carbon dioxide, and water), TBF is resistant to further photooxidation, and its chemical structure suggests resistance to biodegradation.[2]

Methanol has been described as having a high potential for undergoing biodegradation in groundwater after an initial lag period during a controlled release test.[14] The redox conditions for the methanol biodegradation were described as progressing from aerobic to anaerobic to account for the high methanol mass removal. Other studies have also reported biodegradation of methanol.[15,16] The high potential for methanol biodegradation is perhaps owing to the numerous available species of methanogenic bacteria that can utilize methanol as a sole carbon source.[2]

Ethanol, like methanol, is susceptible to aerobic and anaerobic biodegradation.[26] Compared with methanol, the biodegradation of ethanol has been observed as having a longer initial lag period.[15] Ethanol's rate of biodegradation in a gasoline spill would probably be relatively substantial, albeit slower than methanol.[2] However, biodegradation may not be an effective process for relatively high concentrations of methanol and ethanol.[27]

5.6.5 Effects of fuel oxygenates on biodegradation of other gasoline components

The possible inhibitory effects of MTBE and other fuel oxygenates on the biodegradation of fuel hydrocarbons, including BTEX, have been studied, with results that vary.[1] Such inhibition can occur because of effects, toxic to the microcosms, by the inhibitor, or biodegradation of the inhibitor.[2] Hubbard et al.[14] demonstrated no such effect in an aquifer, and Fujiwara et al.[19] found no effect in sludge. However, two studies[20,28] demonstrated inhibitory effects of MTBE on the degradation of BTEX and hexadecane, respectively. Hubbard et al.[14] were able to demonstrate an inhibitory effect of methanol on BTEX degradation. This effect on BTEX was also postulated by Butler et al.[29] for both methanol and ethanol.

5.6.6 Summary of transport and fate of MTBE as a solute in groundwater

Since MTBE and other fuel oxygenates tend to resist adsorption, these chemicals will not be significantly retarded during subsurface transport in a groundwater aquifer unless they are biologically or chemically transformed. BTEX compounds, on the other hand, have a tendency to be retarded owing to both adsorption and degradation mechanisms. Therefore, at the leading edge of a petroleum hydrocarbon plume in groundwater (where levels of the BTEX compounds become non-detectable), MTBE may be found at relatively high concentrations, and persist at a significant distance downgradient from the BTEX component of the plume. And when the fact that MTBE was probably introduced to aquifers later than the initial introduction of BTEX into the aquifer (because the use of MTBE as a gasoline additive is a relatively recent development) is taken into consideration, the relative mobilities of MTBE and BTEX are seen to be even more diverse. More field studies are needed to investigate the effects of transport and fate mechanisms of MTBE in groundwater, as is more evaluation of existing laboratory studies. For more information about MTBE transport in groundwater relative to BTEX transport in groundwater, refer to Appendix H, Plume Geometries for Subsurface Concentrations of MTBE, of this publication.

5.7. Implications of MTBE transport and fate characteristics for environmental restoration projects

Cleanup of groundwater contaminated with MTBE and other fuel oxygenates is difficult due to several qualities of these compounds, including relatively high solubilities and general resistance to biodegradation. Removing MTBE from process water (i.e., extracted groundwater) by aeration methods is difficult owing to the relatively low value of the Henry's law constant for MTBE. Heating water that contains dissolved MTBE will increase the value of the Henry's law constant, making removal easier and perhaps more cost-effective.[30] In contrast, BTEX compounds are more volatile than fuel oxygenates, and therefore easier to remove from water by aeration.[2]

5.7.1 Engineered remedial solutions

Several evaluations of remedial technologies for MTBE in extracted groundwater have indicated generally high costs and low treatment efficiencies. Air stripping has generally been found to have the lowest treatment cost.[31,32] Oxidation of contaminated water using hydrogen peroxide, ozone, or ultraviolet light (UV-Oxidation) is technically feasible but typically expensive.[2] The use of hydrogen peroxide in oxidation also resulted in the breakdown of MTBE and ETBE by hydrolysis, but TBA and acetone were byproducts of the reaction.[17] Breakdown by oxidation of MTBE intermediate products is described in Appendix G, Synthesis, Properties, and Environmental Fate of MTBE and Oxygenate Chemicals. Biotreatment methods may be feasible,[18,33] even though MTBE and other fuel oxygenates are rather resistant to biodegradation, as previously discussed. Treatment by granular activated carbon may not be cost-effective.[11]

A wide variety of in-situ remedial techniques (treatment of the contaminated soil and groundwater in place) are applicable to gasoline hydrocarbons. However, the presence of oxygenates in gasoline-related contamination has caused a reassessment of these techniques as stand-alone remedies for gasoline spill sites. The physicochemical characteristics of oxygenates, previously discussed, are also pertinent in the evaluation of potential in-situ remedial techniques. Air sparging, phytoremediation, enhanced biodegradation, chemical oxidation, reactive barriers, soil vapor extraction, biosparging, and multiphase extraction are the major in-situ techniques that will require more field testing to effectively evaluate their ability to remediate MTBE and other fuel oxygenates in the subsurface.

5.7.2 Natural attenuation

Natural attenuation is currently a common component of cleanup strategies for soil and groundwater contaminated with gasoline hydrocarbons. But the fact that MTBE and some other fuel oxygenates are not strongly retarded (physically or biologically) during their transport in groundwater, as previously discussed, may complicate any cleanup strategy incorporating natural attenuation. Intrinsic bioremediation, however, may be a natural attenuation mechanism that can be realistically applied to the fuel oxygenates methanol and ethanol.[2] Comprehensive field studies that address the intrinsic capability of the subsurface to transform MTBE to nontoxic byproducts are needed. The answer to the question of the best method of remediating fuel oxygenates at gasoline spill sites will undoubtedly be site-specific because of the extreme variability in the hydrogeology, chemistry, and biology of the subsurface.

The following chapter of this publication, Detection and Treatment of MTBE is Soil and Groundwater, explores some of these remedial techniques in greater detail.

Endnotes and references

ROST® is a registered trademark of the Unisys Corporation. Geoprobe® is a registered trademark of Kejr Engineering, Inc.

[1] Squillace, P.J. et al., Review of the environmental behavior and fate of methyl tert-butyl ether, *Environmental Toxicology and Chemistry*, v. 16, n. 9, 1997.

[2] Zogorski, J.S. et al., Fuel oxygenates and water quality, Interagency Assessment of Oxygenated Fuels, The White House Office of Science and Technology Policy, Washington, D.C., July 1997.

[3] Wallington, T.J. et al., Gas-phase reactions of hydroxyl radicals with the fuel additives methyl tert-butyl ether and tert-butyl alcohol over the temperature range 240–440 K, *Environmental Science and Technology*, v. 22, 1988.

[4] Smith, D.F. et al., The photochemistry of methyl tertiary-butyl ether, *International Journal of Chemistry Kinetics*, v. 23, 1991.

[5] Howard, P.H. et al., *Handbook of Environmental Degradation Rates*, Lewis Publishing, Chelsea, Michigan, 1991.

[6] Howard, C.J. et al., Air quality benefits of the winter oxyfuel program, Office of Science and Technology Policy, Washington, D.C., 1996.

[7] Swain, W., Methyl Tertiary-Butyl Ether (MTBE) Overview, U.S. Geological Survey, www.ca.water.usgs.gov/mtbe, 2000.

[8] Schwarzenbach, R.P. et al., *Environmental Organic Chemistry*, John Wiley & Sons, New York, 1993.

[9] Pankow, J.F. et al., Calculated volatilization rates of fuel oxygenate compounds and other gasoline-related compounds from rivers and streams, *Chemosphere*, v. 33, 1996.

[10] Pankow, J.F. et al., The urban atmosphere as a non-point source for the transport of MTBE and other volatile organic compounds (VOCs) to shallow groundwater, *Environmental Science and Technology*, 1997.

[11] Garrett, P. et al., MTBE as a groundwater contaminant., Proceedings of the National Water Well Association/American Petroleum Institute Conference on Petroleum and Organic Chemicals in Groundwater, Houston, Texas, November 12–14, 1986.

[12] Davidson, J.M., Fate and Transport of MTBE—The latest data, Proceedings of the Petroleum Hydrocarbons and Organic Chemicals in Groundwater, Prevention, Detection, and Remediation Conference, Houston, Texas, November 1995.

[13] Prager, J.C., Ed., Methyl tert-butyl ether, *Dangerous Properties of Industrial Materials Report*, v. 12 no. 3, 1992.

[14] Hubbard, C.E. et al., Transport and fate of dissolved methanol, methyl-tertiary-butyl-ether, and monoaromatic hydrocarbons in a shallow sand aquifer, Publication 4601, American Petroleum Institute, Washington, D.C., 1994.

[15] Suflita, J.M. and M.R. Mormile, Anaerobic biodegradation of known and potential gasoline oxygenates in the terrestrial subsurface, *Environmental Science and Technology*, v. 27, 1993.

[16] Mormile, M.R. et al.,. Anaerobic biodegradation of gasoline oxygenates, Extrapolation of information to multiple sites and redox conditions, *Environmental Science and Technology*, v. 28, 1994.

[17] Yeh, C.K. and J.T. Novak, The effect of hydrogen peroxide on the degradation of methyl and ethyl tert-butyl ether in soils, *Water Environmental Research*, v. 67, 1995.

[18] Salanitro, J.P. et al., Isolation of a bacterial culture that degrades methyl t-butyl ether, *Applied Environmental Microbiology*, v. 60, 1994.

[19] Fujiwara, Y. et al., Biodegradation and bioconcentration of alkyl ethers, *Yukayaken*, v. 33, 1984.

[20] Jensen, H.M. and E. Arvin, Solubility and degradability of the gasoline additive MTBE, methyl-tert-butyl-ether and gasoline compounds in water, F. Arendt, M. Hinsenveld, and W.J. van den Brink, Eds., *Contaminated Soil*, Kluwer, Dordrecht, The Netherlands, 1990.

[21] Yeh, C.K. and J.T. Novak, Anaerobic biodegradation of oxygenates in the subsurface, from Proceedings, Petroleum Hydrocarbons and Organic Chemicals in Groundwater, Prevention, Detection and Restoration, Houston, Texas, November 20-22, 1991.

[22] Yeh, C.K. and J.T. Novak, Anaerobic biodegradation of gasoline oxygenates in soils, *Water Environmental Research*, v. 66, 1994.

[23] Daniel, R.A., Intrinsic bioremediation of BTEX and MTBE: Field, laboratory, and computer modeling studies, M.S. thesis, North Carolina State University, Raleigh, North Carolina, 1995.

[24] Thomas, J.M. et al., Environmental fate and attenuation of gasoline components in the subsurface, Final Report, American Petroleum Institute, Washington, D.C., 1998.

[25] Hickman, G.T., Effects of site variations on subsurface biodegradation potential, *Journal of Water Pollution Control*, v. 61. February, 1989.

[26] Chapelle, F.H., *Groundwater Microbiology and Geochemistry*, John Wiley & Sons, New York, 1992.

[27] Brusseau, M.L, Complex mixtures and groundwater quality, U.S. Environmental Protection Agency, Robert S. Kerr Environmental Research Laboratory, EPA/600/S-93/004, Washington, D.C., 1993.

[28] Horan, C.M. and E.J. Brown, Biodegradation and inhibitory effects of methyl-tertiary-butyl ether (MTBE) added to microbial consortia, Proceedings, 10th Annual Conference on Hazardous Waste Research, Manhattan, KS, May 23–25, 1995.

[29] Butler, B.J. et al., Impact of methanol on the biodegradation activity of aquifer microorganisms. Presented at SETAC, 13th Annual Meeting, November 8–12, Cincinnati, Ohio, 1992.

[30] Butillo, J.V. et al., Removal efficiency of MTBE in water: Confirmation of a predictive model through applied technology, Proceedings of the Petroleum Hydrocarbons and Organic Chemicals in Groundwater, Dublin, Ohio, November 2–4, 1994.

[31] International Technology Corporation, Cost-effective, alternative treatment technologies for reducing the concentrations of ethers and alcohols in groundwater, Publication 4497, American Petroleum Institute, Washington, D.C., 1991.

[32] Truong, K.N. and C.S. Parmele, Cost-effective alternative treatment technologies for reducing the concentrations of methyl tertiary butyl ether and methanol in groundwater, *Hydrocarbon Contaminated Soils and Groundwater*, v. 2, 1992.

[33] Mo, K., C.O. Lora, A. Wanken, and C.F. Kulpa, Biodegradation of methyl-t-butyl ether by pure bacterial cultures, Abstracts, 95th General Meeting of the American Society for Microbiology, Washington, D.C., May 21–25, 1995.

chapter six

Detection and treatment of MTBE in soil and groundwater

6.1 Introduction

In contrast to gasoline additives used in the past, methyl *tertiary*-butyl ether (MTBE) is in a different class of chemical compounds: ethers. Ethers possess unique properties — namely, enhanced solubility in water and chemical attraction to water molecules — which increase the potential for environmental problems associated with gasoline leaks and spills.

The following discussion focuses on the current status of technology for detecting and quantifying the presence of MTBE in groundwater and soil, and methods for restoration of these resources should they become contaminated with MTBE.

6.2 Analysis of samples for detection of MTBE

The detection and quantitative measurement of MTBE in environmental media, such as soil and groundwater, have not been exhaustively studied. In fact, there are no accepted measurement methods promulgated by regulatory agencies addressing the compound, nor do any similar compounds, such as TAME or ETBE, have promulgated analysis methods. The future holds promise for issuance of such methods. At the present time advanced analytical methodology is available that will provide the necessary analyses; the user of those methods is cautioned, however, regarding the above conditions. Aspects of the currently available methodology are discussed below.

6.2.1 Analytical methodology: validation

Samples taken for environmental compliance or assessment purposes are required to meet standards of precision and accuracy established by regulatory agencies. Of particular note in this regard is the United States Environmental Protection Agency (U.S. EPA)'s promulgation of enforceable analysis procedures as described in Section 304(h) of the Clean Water Act (CWA). Methods to be promulgated for enforcement purposes under that section of the CWA have been adapted for use under other environmental statutes such as the Resource Conservation and Recovery Act (RCRA) and the Safe Drinking Water (SDW) Act.

Methods promulgated under Section 304(h) have documented sensitivity and accuracy through application of the procedures identified in 40 CFR Part 136, Appendix B, a process generally known as method validation. The process includes determining the performance and capability of the method from the preparation procedure to detection and quantitative analysis. For example, in validation of U.S. EPA Method 602 (addressing the group of chemicals known as aromatic hydrocarbons), the performance of each of the individual elements of the method (e.g., the purge and trap procedure, the column used to separate the components, and the detector used to quantify their presence) was evaluated. The overall performance of the method is defined by confirmation of the presence of the analyte, or chemical tested, and its quantitative analysis in the sample. Given that the compound must be detected with 100% confidence, its recovery from the sample matrix must lie within a reliable range to conclude that the method is valid for the analyte. The validation procedure must be carried out for each sample media of interest, i.e., solid waste, waste water, air, etc.

Additional studies are also made to determine if the validity of the method is equivalent among the many laboratories that will be performing the analysis. This level of validation ensures that accurate data are available nationwide.

6.3 *Validated methods for detection of MTBE in drinking water*

In the case of MTBE, a relatively new compound of regulatory interest, only two validated methods established for drinking water are available: U.S. EPA Methods 524.2 (Rev. 3) and 502.2. U.S. EPA Method 524.2 utilize the standard chromatography column for separation of the target analytes and detection with mass spectrometry. The target analyte list includes MTBE. Similarly, U.S. EPA Method 502.2 uses a standard volatiles column for separation of the analytes and detection with a photoionization detector. These methods are probably the easiest to validate since drinking water contains few background compounds that would potentially interfere with the analysis. In the absence of other acceptable analyses, these methods may be suitable.

6.4 *Nonvalidated methods for detection of MTBE in drinking water*

A number of alternative methods exist for detection of MTBE in water samples. U.S. EPA Method 8260, a method developed for solid waste, is the functional equivalent of 524.2 and has been recommended for use in the detection and quantitative measurement of MTBE. MTBE was not considered a candidate analyte during the establishment of this analytical method; thus, this method has not been validated for use in detection and quantification of MTBE.

Studies are currently in progress for validating U.S. EPA Method 8260 for use in analyses of samples for MTBE.

The most common method for analysis of samples for total petroleum hydrocarbons and the aromatics benzene, toluene, ethylbenzene, and xylenes (BTEX) is U.S. EPA Method 8020. In the common usage of this method, a sample of water containing hydrocarbons is purged and trapped, and the compounds on the trap are desorbed onto a standard volatiles column. The compounds eluted from the column are detected first by a photoionization detector (PID) and then a flame ionization detector (FID). The PID is sensitive to aromatics and selected aliphatic hydrocarbons, and the FID detects all hydrocarbons. MTBE responds to both the PID and FID.

Complications occasionally arise in the analyses, however, since the common volatiles column does not provide separation of the methyl pentanes and MTBE. Unfortunately, the methyl pentanes also respond to the PID, causing potential interferences with the detection of MTBE. If MTBE is present with hydrocarbons, continued analysis by U.S. EPA Method 8260 may be required. If the samples do not contain elevated concentrations of hydrocarbons, U.S. EPA Method 8020 can be used to detect and quantify the presence of MTBE.

A third option is available. Selected chromatography columns are available that will separate the methyl pentanes from MTBE allowing, in some cases, the use of U.S. EPA Method 8020. Inquiries must be made of the project laboratory to determine the availability of selective columns for the analyses required.

A review of analysis methodology, conducted by Lawrence Livermore National Laboratory (LLNL), is provided below.

> We have tested analytical methods to determine the precision and accuracy of oxygenate analysis in groundwater containing dissolved gasoline compounds. EPA Method 8260A and a modified version of ASTM Method D4815 produced excellent results for all analytes regardless of the amount of gasoline interfaces present in the sample.
>
> Overall, EPA Method 8020A/21B was identified as a very conservative monitoring tool due to the lack of false-negative results, and its tendency for over-estimation of analyte concentrations and false-positive misidentifications. However, in the absence of required method modifications, more definitive tests such as EPA Method 8260A and the modified ASTM Method D4815 are recommended when monitoring low concentrations of oxygenates in samples that may have high regulatory impact.[1]

6.5 Methods of analysis for detection of MTBE in other media

Methods for detecting and quantifying MTBE in air, body fluids, ecological media, and soil vapor are also available. Since such methodologies have not been through any validation procedure sponsored by a regulatory agency, the burden of validation is on the investigator to demonstrate the performance of the method in the sample matrix of interest. Each of these media and the reported methods are discussed below.

6.5.1 Analysis of samples in ambient air

In a recent study to determine ambient air concentrations of MTBE in California cities by the California Air Resources Board, a method was used where samples were taken in Suma canisters and analyzed by a procedure called two-dimensional gas chromatography.[2] The analysis protocol involved three steps: (1) the gas sample (150 cm^3) was preconcentrated in a cryogenically cooled trap to eliminate air as the sample matrix; (2) the sample was then thermally desorbed onto a DB-Wax capillary pre-column; and (3) the compounds of interest were chromatographed through a DB-1 column to a FID.

The use of the pre-column removes water vapor from the sample which would interfere with the detection of MTBE. After analysis of a sample, water and other compounds are allowed to vent to the atmosphere prior to analysis of the next sample. The reported detection limit for MTBE using this method is 0.2 parts per billion by volume (ppbv). The detection limit of 0.2 ppbv would be suitable for tracking an MTBE plume.

6.5.2 Analysis of body fluids

Studies have been made of the presence of MTBE in blood and urine.[3] Two methods — purge and trap GC–MS, and purge and trap GC–M using an isotope dilution procedure — have been used.

The methods are similar in that the purge and trap procedure was used to remove MTBE from the fluid under analysis, and the columns and detector were identical. The use of isotope dilution requires the addition of a "cocktail" of the isotopically labeled compounds of interest to the sample prior to purging. The labels are normally deuterium (^{2}H) instead of hydrogen, or ^{13}C rather than ^{12}C.

The use of a mass spectrometer as the detector for the analysis allows easy detection of the labeled compounds and calculation of their recovery from the purged sample. Knowing the quantity of the labeled compound added to the sample and the quantity measured in the analysis allows calculation of recovery and, thus, determination of the performance of the method for the matrix under evaluation.

Excellent results were obtained for both of the methods.

6.5.3 Analysis of soil vapor samples

The use of soil vapor surveys as an assessment procedure is well-recognized by members of the soil and groundwater remediation industry. Analysis of the soil vapor samples poses the same problems with chromatographic interferences from the methyl pentanes, as discussed above for drinking water samples. Analysis performed through the use of soil vapor surveys, however, has some inherent advantages over analysis of water samples, in that the concentration of MTBE in gasoline is substantially greater (in the range of 10 to 15%), than the methyl pentanes (1 to 3%). The concentration difference helps to offset the difference in the Henry's law constant between the hydrocarbons and MTBE.

As noted above in the analysis of samples in ambient air by the two-dimensional gas chromatographic method, the detection limit of 0.2 ppbv is considered suitable for tracking an MTBE plume. The equilibrium relationship between the vapor and groundwater concentrations of MTBE are shown in Table 6.1 for a Henry's law constant of 0.026 at 25 C.

Table 6.1 Estimated Concentration of MTBE in Soil Vapor Relative to Water

Concentration in Water μg/L	Concentration in Soil Vapor ppm(v)	Concentration in Soil Vapor μg/m^3
1	0.00662	0.026
5	0.0331	0.13
10	0.0662	0.26
50	0.331	1.3
100	0.662	2.6
500	3.309	13
1,000	6.618	26
5,000	33.091	130
10,000	66.182	260
43,000	284.582	1118

As can be seen from the above table, a quantification limit of 0.2 ppbv will allow tracking of an MTBE plume to very low concentrations if the equilibrium concentrations are achieved in the soil vapor column. The extent and accuracy to which a groundwater plume of MTBE can be tracked with soil vapor surveys have not yet been determined.

6.6 Analysis of other gasoline additives

Since groundwater may contain an unknown variety of gasoline additives and oxygenates, analysis of groundwater must include not only MTBE, but other chemicals as well. MTBE is typically analyzed using U.S. EPA Method 8020, which is the same method used to analyze the concentrations of BTEX present in a sample. The significant problem with Method 8020 is that multiple analytes can co-elute from the column; for example, TAME may co-elute with benzene in Method 8020. Gas Chromatography (GC)-Mass Spectrometry (MS) U.S. EPA Method 8260 is a more definitive procedure to determine oxygenate compounds. GC-MS will likely be more costly, and may not detect methyl and ethyl alcohols very well, but this method is still the best analytical method available at present.[4]

6.7 Variations in analysis resulting from presence of BTEX constituents

As long as it is present at concentrations similar to the BTEX compounds, MTBE can be quantitatively analyzed with one analytical test along with the BTEX compounds. However, where MTBE concentrations are substantially higher than the associated BTEX compounds, the BTEX compounds produce responses in the middle of their calibration ranges, and MTBE will often far exceed its upper calibration limit to the point of saturating the detector. This occurs because MTBE's solubility in water is greater than that of the BTEX compounds. MTBE is 25 times more soluble than benzene and 70 times more soluble than toluene. Laboratories may be required to run the groundwater sample analysis twice to obtain quantitative results for MTBE because the samples will have to be diluted to bring MTBE within the calibration range. This is especially true for samples taken from downgradient groundwater monitoring wells where samples frequently contain higher concentrations of MTBE and low or no concentrations of the BTEX compounds.[5]

6.8 Treatment of MTBE in groundwater

The high water solubility and resistance to biodegredation of MTBE make remediation of contaminated groundwater expensive and generally inefficient. Numerous studies have been performed to evaluate the cost and effectiveness of treatment technologies for removal of MTBE from groundwater. The results of these studies can be generally summarized as follows:

- Although MTBE can be removed from groundwater by physical technologies such as activated carbon adsorption and air stripping, the cost-effectiveness of these technologies in removal of MTBE is approximately 10 times less than their application for removal of hydrocarbons, such as benzene and toluene, from groundwater.
- Although MTBE can be metabolized by acclimated bacteria, resulting in its mineralization (conversion to carbon dioxide), the rate of growth of the bacteria is quite slow. There are no reports assessing the field applicability of this technology.
- Chemical oxidation of MTBE has been successful in the laboratory. The use of Fenton's reagent (H_2O_2 and Fe(II)) has great potential in successfully treating MTBE and BTEX in groundwater.

The major reason MTBE is a significant cleanup concern is that its physical properties make conventional cleanup technologies extremely costly and relatively inefficient. Organic contaminants in soil and groundwater can be analyzed according to their solubility, volatility, and density with respect to water. MTBE is less dense than water, has a high solubility, and has a low partitioning constant for sorption onto organic matter in soil. MTBE also has low volatility and a low Henry's law constant.

Because of MTBE's chemical and physical properties, there is a low efficiency of MTBE removal from groundwater using conventional cleanup technologies. Adsorption onto carbon is low (maximum mass loadings of 1 to 3% vs. approximately 20% for benzene) because of the high solubility and low partitioning constant of MTBE. Air-sparging as a cleanup strategy for MTBE is also inefficient, owing to MTBE's low volatility and low Henry's law constant. Recent data suggest that magnesium peroxide (sometimes called Oxygen Release Compound, or ORC®), is not an effective cleanup tool for MTBE on a widespread basis. Aboveground microbial reactions have been shown to work effectively, but these reactions are difficult to control and maintain. Therefore, a need exists to identify more effective tools to treat MTBE in the subsurface.

Details of the literature reports of findings about various methods of MTBE remediation are presented in the following discussion. The reader is cautioned that the sources of the details of these technologies are often materials prepared by vendors or proponents of these technologies.

6.8.1 *Adsorption by activated carbon*

Two studies addressing the removal of MTBE from water through adsorption on activated carbon have been completed.[6] In the study from the Oxygenated Fuels Association (OFA), an assumed concentration of 700 ppbv of MTBE was used to calculate the cost of treatment of an effluent of 35 ppb at two flow rates (6000 gallons per minute and 600 gallons per minute). The costs were calculated assuming a 20-year period at a 4% discount rate. The results are presented in Table 6.2.

Table 6.2 Costs of MTBE Removal by Activated Carbon

Flow Rate	Cost	Capital Cost
6,000 gpm	$1.96/1,000 gallons	$2,400,000
600 gpm	$2.03/1,000 gallons	$378,000

(From: OFA, 1997.)

In the study from the American Petroleum Institute (API), activated carbon was used as a polishing technology for the effluent from an air stripper. The assumed concentration of MTBE into the stripper was 20 parts per million (ppm), and the flow rate was 25 gallons per minute. The effluent from the stripper and influent to the carbon was calculated to contain 500 ppb of MTBE. The effluent criterion for MTBE was 10 ppb. Using these criteria, the cost of combined treatment was calculated to be $10 per 1000 gallons of water. Allocation of the cost specific to the air stripper vs. the carbon adsorber is not possible with the data provided.

An assessment of the effectiveness of activated carbon indicated that the efficiency of carbon for hydrocarbon removal from groundwater was 2.0%, and that the efficiency of carbon for MTBE removal from groundwater was less than 0.3%.[8] Efficiency is defined as the percent by mass of the compound adsorbed to carbon. An earlier study of 15 air-stripping installations indicated that 56 to 99.9% of the MTBE had been removed from the source tested. The median removal was 91%.[7]

6.8.2 *Air stripping*

Two studies have been made which address the cost and effectiveness of air stripping of MTBE from water. One study was sponsored by the OFA. A second study compared the air : water ratios necessary for removal of petroleum hydrocarbons to the ratio for removal of MTBE.[10] The results of the OFA's evaluation showed that, for comparable reductions, the air : water ratio for MTBE was approximately 7 times greater than that for hydrocarbons; specifically, the ratio was 1000 : 150.

The OFA study data were generated for the same flow rates as the activated carbon evaluation. The resulting data are provided in Table 6.3.

Table 6.3 Costs of Air Stripping for Removal of MTBE from Water

Flow Rate	Cost	Comments
6,000 gpm	$0.21/1,000 gallons	no off gas treatment
6,000 gpm	$0.55/1,000 gallons	with off gas treatment
600 gpm	$0.47/1,000 gallons	no off gas treatment
600 gpm	$0.83/1,000 gallons	with off gas treatment

(From: OFA, 1997.)

6.8.3 *Biological treatment*

The biodegradation of MTBE has been attempted in numerous studies. While evidence exists that MTBE degrades biologically at some sites contaminated by gasoline, successful degradation has been elusive owing to the nature of the intermediate compound, *tertiary*-butyl alcohol (TBA), which is formed in the process along with methanol. For some time, TBA has been described as a compound that was slowly degraded, if at all, in mixed culture systems.

Details from three studies describe the circumstances in which biotreatment has been evaluated.[9–12] The study conducted by Fujiwara et al. evaluated the degradation of MTBE as a co-metabolite in the degradation of the hydrocarbons found in gasoline. The study results indicated that although the gasoline hydrocarbons were completely degraded, MTBE was not degraded. This study also concluded that MTBE did not interfere with the degradation of gasoline components.

The study by Yeh and Novak[9] found no degradation of TBA, formed after treatment of MTBE with hydrogen peroxide.

In contrast to these two earlier studies, the experiments performed by Mosteller[16] and Cowan and Park[12] were successful in degrading MTBE. The study by Mosteller was carried out using a fluidized bed bioreactor. In that device, MTBE is removed through adsorption to the surface, or within micropores of the fluidizing material, after which the bacteria attached to the material metabolize the adsorbed MTBE. This process is commonly used in the removal of recalcitrant organic compounds in industrial effluents, such as those from the petroleum refining industry. After acclimation, the results of the study indicated a 75% reduction in the concentration of MTBE across the reactor.

In the Cowan and Park study,[17] MTBE was degraded to carbon dioxide through the use of acclimated bacteria. The specific substrate parameters for removal are provided in Table 6.4 as a function of temperature. (In the table, u_{max} represents the rate of degradation.)

Table 6.4 Biodegradation Rate of MTBE as a Function of Temperature

Temperature (°C)	Initial Concentration (ppm)	u_{max} (hr^{-1})
20	100	0.012
25	100	0.025
30	100	0.043

(Source: Park et al., 1997.)

In these experiments the dissolved oxygen concentration was found to be a critical factor. Concentrations from 3.6 to 7.4 ppm were essential for bacterial metabolism of MTBE.

6.8.4 Chemical oxidation

Chemical oxidation has been evaluated as a treatment for MTBE in groundwater. In studies by Sevilla et al.[7] and the OFA,[6] experiments were carried out to determine the cost effectiveness of treatment. Table 6.5 provides the results of those studies.

Table 6.5 Costs of Chemical Oxidation for Removal of MTBE from Water

Treatment Type	Cost or Cost Effectiveness
Advanced Oxidation, 6,000 gpm	$0.39-0.44/1,000 gallons
Advanced Oxidation, 600 gpm	$0.50-0.55/1,000 gallons

(From McGrath, A., SECOR International, Inc., 1999. With permission.)

Chang and Young[13] evaluated chlorine, ultraviolet (UV) light, and UV/hydrogen peroxide (UV/H_2O_2) treatments. Their results indicated that chlorine had no effect on MTBE. UV and UV/H_2O_2 treatments had higher efficiencies in degrading MTBE, but both generated *tertiary*-butyl formate (TBF) as the major byproduct (at a concentration of approximately 0.1% minimum TBF). The reaction was 99.9% efficient at a concentration of 10 milligrams per liter (mg/L). This represents a residual MTBE concentration of approximately 10 mg/L, with a residual TBF concentration greater than 150 mg/L.

6.8.5 *Chemical oxidation using Fenton's reagent*

Researchers analyzing the efficiency of a chemical oxidant known as Fenton's reagent (hydrogen peroxide and ferrous iron, Fe(II) at pH <4.5) found that the only major by-products of the reaction were TBA and acetone at very low concentrations. Fenton's reagent is potentially an attractive treatment because it may be used to treat groundwater in situ, depending on site conditions. The following section provides an overview of proprietary laboratory studies conducted to evaluate the effects of groundwater pH and soil colloids on oxidation of MTBE by Fenton's reagent.

6.8.6 *Overview of Fenton's reagent*

Fenton's reagent was developed in the late 1890s when it was discovered that iron, acting as a catalyst, enhanced the reactions of hydrogen peroxide. Fenton's reagent is known to oxidize and, in some cases, completely mineralize a variety of organic substrates. Although the reaction mechanism is not completely understood, it is generally accepted that Fenton's reagent generates hydroxyl free radicals (OH·) that in turn react with available organic substrates (R) which undergo oxidation.

H_2O_2 + Fe(II) fi OH^- + OH· + Fe(III)

OH· + RH fi R· + H_2O

C_6H_6 (benzene) + 15 H_2O_2 [Fe(II) catalyst] fi 6 CO_2 + 18 H_2O

A variety of other competing reactions consume the hydroxyl free radical, making dosing predictions difficult. However, published studies have documented the ability of Fenton's reagent to degrade many types of organic environmental contaminants, including the BTEX compounds, MTBE, trichloroethene (TCE), tetrachloroethene (PCE), pentachlorophenol (PCP), nitrobenzene, and many other chlorinated hydrocarbons that are otherwise resistant to degradation. Substances such as the BTEX compounds have shown particular susceptibility to oxidation by Fenton's reagent. These compounds degrade rapidly and can be completely converted to carbon dioxide. Incomplete degradation products include carboxylic acids and alcohols, which are rapidly degraded by microbes, and are less toxic than the BTEX compounds.

The oxidation of MTBE can result in the formation of a series of degradation products, including TBF and TBA. TBF and TBA form through the following reactions:

$(CH_3)_3\text{-C-O-CH}_3 + 3\ H_2O_2$ [Fe(II) catalyst] fi

$(CH_3)_3\text{-C-O-COOH (TBF)} + 4\ H_2O$

$(CH_3)_3\text{-C-O-CH}_3 + 3\ H_2O_2$ [Fe(II) catalyst] fi

$(CH_3)_3\text{-C-OH (TBA)} + CO_2 + 4\ H_2O$

One challenge in applying Fenton's reagent in situ is that groundwater concentrations of ferrous iron are typically low, and the pH is not low enough in soils and groundwater to maintain the iron required to catalyze the reaction in solution. Dissolved iron precipitates out of solution at $pH > 5$ and cannot catalyze the Fenton's reaction to generate the hydroxyl radical. Some researchers have demonstrated that metal ligands (chelating agents) such as nitrolotriacetic acid (NTA) and gallic acid dissolve iron at $pH > 5$, and allow for the formation of the hydroxyl radical. Geochemistry studies conducted using several bench-scale tests in the laboratory have confirmed the ability of Fenton's reagent to treat BTEX compounds and MTBE. Furthermore, field tests conducted at contaminated sites have suggested that in-situ application of Fenton's reagent has been used to successfully treat BTEX, solvents, diesel, and gasoline in contaminated groundwater.[14,15] In addition, treatment with a metal chelating agent improves the reaction of Fenton's reagent at $pH > 5$, as is shown in the bench-scale test results shown in Table 6.6. Complexing agents are not required when the soil buffer capacity indicates that pH adjustments are possible.

6.8.7 *Fenton's reagent bench-scale test*

Bench-scale tests were designed and conducted to determine the most effective dosing for in-situ treatment with Fenton's reagent. The bench-scale tests

Table 6.6 Laboratory Bench-Scale Test Results for BTEX Oxidation by Fenton's Reagent

Treatment	Benzene*	Toluene*	Ethyl-Benzene*	Xylenes*	MTBE*	TBA*	TBF*	TPH*
Control	930	14,000	2,400	14,000	30,000	<5,000	<500	96,000
1.0% H_2O_2	<0.5	<0.5	<0.5	<0.5	<0.5	<5.0	<0.5	<50
2.0% H_2O_2	<0.5	<0.5	<0.5	<0.5	<0.5	<5.0	4.2	<50
3.0% H_2O_2	<0.5	<0.5	<0.5	<0.5	<0.5	<5.0	<0.5	<50
3.0% H_2O_2†	<0.5	<0.5	<0.5	<0.5	1	<5.0	320	<50

* μg/L † no pH adjustment, pH > 6

(From McGrath, A., unpublished case studies, SECOR, 1999. With permission.)

evaluated the chemistry of the introduction of Fenton's reagent to sediments with a moderate particle size and buffer capacity. The reactions included five 250-mL reaction vessels that were prepared by mixing 100 mL of soil, 50 mL of groundwater, and 250 mL of gasoline containing MTBE. Two different treatments were tested:

- Fenton's reagent with acidification.
- Fenton's reagent without acidification (no pH adjustment).

Peroxide concentrations were set at 0 (for control sample), 1, 2, and 4% using a 35% peroxide concentrate solution. Iron concentrations were set at 250 mg/L Fe(II) for each 50 mL of groundwater added to soil. The pH of the Fenton's reagent sample was acidified to pH between 3.5 and 4 prior to Fe(II) addition and peroxide addition. The control and no acidification samples had no pH adjustment (pH ~ 6 to 7).

Results of the bench-scale test indicate that BTEX and MTBE are degraded effectively by Fenton's reagent with and without acidification. The sample without pH adjustment did contain some residual MTBE and degradation products, but treatment was highly effective.

6.8.9 Conclusions of tests using Fenton's reagent

Based on the results of the bench-scale studies conducted, Fenton's reagent has great potential for treating MTBE and BTEX in groundwater. The limits of in-situ MTBE remediation will be affected by the permeability of the aquifer and alkalinity of the groundwater and soils: the higher the pH and alkalinity, the lower the reactivity of Fenton's reagent. Treatment is most effective in high permeability, low pH (pH of between 2 and 4) aquifers having low alkalinity.

6.8.10 Hydrogen peroxide bench-scale tests

In a different set of experiments, analyses were performed in glass beakers in an analytical laboratory. In 4- and 24-hour experiments, equal amounts of

Table 6.7 Laboratory Bench-Scale Test Results for MTBE Oxidation by Hydrogen Peroxide (20%)

Phase I – 4-hour Experiment	Results
MTBE Concentration	1,086 mg/L
Hydrogen Peroxide Concentration	20 percent
Mixed MTBE Concentration	543 mg/L
Concentration after 4 hrs. with mixing*	48.65 mg/L
Percent Reacted in 4 hrs.	91 percent reduction
Phase II – 24-hour Experiment	
MTBE Concentration	981 mg/L
Hydrogen Peroxide Concentration	20 percent
Mixed MTBE Concentration	491 mg/L
Concentration after 24 hrs. with mixing*	<0.5 μg/L
Percent Reacted in 24 hrs.	99.9** percent reduction

*US EPA Method 8020
**Five unidentified degradation products were detected at a concentration for all five products of less than 0.05 ppm
(Source: Fast-Tek, 1999. With permission.)

MTBE were added to hydrogen peroxide at 20% concentration. During the degradation of MTBE by hydrogen peroxide, some breakdown compounds were formed. Some of the more stable daughter products of this reaction are glycolic acid and hydroxyl acids, formaldehyde, and some alcohols such as *tertiary*-butyl alcohol (TBA). The results of these experiments are presented in Table 6.7.

Experiments assessing the effectiveness of hydrogen peroxide in degrading MTBE are still ongoing. It appears that the end products of the reaction of MTBE with hydrogen peroxide are carbon dioxide and water. High-pressure (3000 to 5000 pounds per square inch) injection of hydrogen peroxide or other oxidants into the subsurface requires closely spaced injection points (2 to 5 ft centers) to provide effective coverage and mixing of the oxidation chemicals with MTBE. Nonetheless, treatment of MTBE must be evaluated on a site-by-site basis.

6.8.11 Discussion & conclusions

The results of the treatability studies indicate that MTBE is difficult and costly to remove from groundwater, when compared to the traditional treatment costs and difficulties experienced with the removal of hydrocarbons found in gasoline, such as benzene. If cost were no object, there appears to be potential for use of oxidants, air strippers, and carbon adsorption for aboveground treatment.

Full-scale implementation of biological treatment alternatives has not been reported, although experiments with the fluidized bed bioreactor are apparently in progress. The substrate removal rate for MTBE in these experiments with the fluidized bed reactor is low in comparison to other hydrocarbons. Special reactor design and extended holding times will likely be required for complete removal.

Various populations of MTBE-degrading microorganisms have been successful in degrading MTBE in the laboratory;[16–19] however, there is little evidence that these destructive processes occur quickly or commonly in field conditions.[1] Nonetheless, successful biotreatment using engineered bioremediation techniques may potentially be developed.

Based on the results of the bench-scale studies conducted, Fenton's reagent has great potential for treating MTBE and BTEX in groundwater. The limits of in-situ MTBE remediation by this method will be affected by the permeability of the aquifer and alkalinity of the groundwater and soils: the higher the pH and alkalinity, the lower the reactivity of Fenton's reagent. Treatment is most effective in high permeability, low pH ($pH < 7$) aquifers where injection of peroxide (or potassium permanganate, another commonly used oxidant) is relatively easy and the pH may be acidified with minimal effort.

Endnotes and references

ORC® is a registered trademark of the Regenesis Bioremediation Products, Inc.

[1] Lawrence Livermore National Laboratory (LLNL), An Evaluation of MTBE Impacts to California Groundwater Resources, UCRL-AR-130897, 2000.

[2] Poore, M. et al., Sampling and analysis of t-butyl ether in ambient air at selected locations in California, Proceedings of the 213 Meeting of the *ACS*, San Francisco, California, 1997.

[3] Bonin, M.A. et al., Measurement of methyl t-butyl ether and t-butyl alcohol in human blood and urine by purge and trap gas chromatography-mass spectrometry using an isotope dilution method, American Chemical Society Division of Environmental Chemistry pre-prints of papers, 208th, Washington, D.C., v. 34, n. 2, 1997.

[4] Boggs, G.L., Analysis Required for Oxygenate Compounds Used in California Gasoline - EPA Method 8260 (8240-B and 8020), open file letter, California Regional Water Quality Control Board, Central Valley Region, August 30, 1997.

[5] Sequoia Analytical, MTBE: Solubility in water and sample analysis, *Sequoia Sentinel*, v. 2.1, Winter, 1996.

[6] Oxygenated Fuels Association (OFA), Estimates of Annual Costs to Remove MTBE from Water for Potable Uses, Technical Memorandum, 3195-001, May, 1997.

[7] Sevilla, A. et al., Effect of MTBE on the Treatability of Hydrocarbons in Water, Environmental Science and Technology Chemical Preprints of Abstracts, v. 37, no. 1, April 1997.
[8] Groundwater Technology, Inc., A Compilation of Field-Collected Cost and Treatment Effectiveness Data for the Removal of Dissolved Gasoline Components from Groundwater, American Petroleum Institute Publication No. 4525, November, 1990.
[9] Yeh, C. K. and Novak, J. T., The effect of hydrogen peroxide on the degradation of methyl and ethyl *tert*-butyl ether in soils, *Water Environ. Res.*, v. 57, no. 5, July/August, 1995.
[10] Fujiwara, Y., Biodegradation and bioconcentration of alkyl ethers, *Yukayaken*, v. 33, 1984.
[11] Park, K. et al., Effects of oxygen and temperature on the biodegradation of MTBE, *Environ. Sc. Technol.*, Preprints of Extended Abstracts, v. 37, no. 1, April, 1997.
[12] Mosteller, D. C. et al., Biotreatment of MTBE Contaminated Groundwater, *Environ. Sci. Technol.*, Preprints of Extended Abstracts, v. 37, no. 1, April, 1997.
[13] Chang, C. Y. and Young, W. F., *Health and Environmental Assessment of MTBE: Report to the Governor and Legislature of the State of California as Sponsored by SB 521*, Volume V: Risk Assessment, Exposure Assessment, Water Treatment and Cost-Benefit Analysis: Reactivity and By-Products of MTBE Resulting from Water Treatment Processes, University of California, November 12, 1998.
[14] McGrath, A., Unpublished Case Studies, SECOR International, Inc., Oakland, California, 1999.
[15] Jacobs, J., Unpublished Case Studies, FAST-TEK Engineering Support Services, Pt. Richmond, California, 1999.
[16] Salanitro, J.P. et al., Isolation of a bacterial culture that degrades methyl tert-butyl ether, *Appl. Environ. Microbiol.*, v. 60 n. 6:2593–2596, 1994.
[17] Cowan, R. and Park, K., Biodegradation of Gasoline Oxygenates MTBE, ETBE, TAME, and TBA by an Aerobic Mixed Culture, Proceedings of the 28th Mid-Atlantic Industrial and Hazardous Waste Conference, Buffalo, New York, pp. 523–530, 1996.
[18] Mo, K. et al., Biodegradation of Methyl Tert-Butyl Ether by Pure Bacterial Cultures, *Appl. Microbiol. Biotech.*, v. 47:69-72, 1997.
[19] Steffan, R.J. et al., Biodegradation of the gasoline Oxygenates MTBE, ETBE and TAME by Propane-Oxidizing Bacteria, *Applied Envir. Microbiology*, v. 63(11):4216-4222, 1997.

chapter seven

MTBE: A perspective on environmental policy

7.1 Introduction

In March 2000, the United States Environmental Protection Agency (U.S. EPA) issued a recommendation to discontinue addition of the gasoline additive MTBE to fuel refined and sold in the U.S. The U.S. EPA recommendation came 21 years after the original introduction of MTBE as a fuel oxygenate and octane booster for gasoline, and marked an important milestone in the increasingly vocal public debate over health hazards and groundwater contamination that are associated with use of the compound.

MTBE was introduced to the market under patent by the Atlantic Richfield Company (ARCO) in 1979. In light of the continuing controversy over health effects associated with the accidental ingestion of MTBE, the rise of MTBE's use is strikingly ironic, as the additive was developed primarily as an octane booster to enhance the performance of unleaded fuels — fuels introduced in response to health and environmental concerns posed by leaded gasoline.

The 1990 Amendments to the federal Clean Air Act (1990 CAA) mandated the addition of oxygenates to gasoline in regions that failed to meet federal air quality standards. MTBE and ethanol are currently the two gasoline oxygenates most commonly used in the U.S. to reduce carbon monoxide (CO) and ozone (O_3) motor vehicle emissions, though MTBE is by far the more commonly used additive. U.S. production of MTBE in 1995 totaled 8.0 billion kilograms, roughly double that of ethanol in the previous year.[1] Because MTBE not only reduces CO emissions by approximately 10%, but also serves as an octane enhancer in gasoline, the additive seemed to provide an answer to the oil industry's drive toward better fuel efficiency and environmentalists' concerns about increasingly disastrous air quality standards in major urban areas. The fact that MTBE is manufactured from chemicals that were previously refinery byproducts made it even more attractive to the petrochemical industry.

After its widespread introduction via oxygenated fuel programs in the early 1990s, however, MTBE was discovered to be, for all its benefits, yet another challenge for petroleum distributors and environmental quality regulators alike. The positive air quality benefits of MTBE have become overshadowed by the public reaction to the use of MTBE in gasoline. The most persistent and potentially harmful environmental side effect of the widespread use of MTBE is its appearance in groundwater and drinking water wells, which poses a threat to the beneficial uses of potable aquifers across the U.S. In part because of the complex nature of MTBE, and in part because of the varied public reaction MTBE has inspired, policy that directly regulates this fuel oxygenate is often fractured in its application, differs widely from state to state, and has not always been consistent with the U.S. EPA's drinking water advisory for MTBE.

This chapter analyzes the manner of MTBE regulation in California and the U.S., and evaluates the environmental policy-making process as it pertains to MTBE. The results of this analysis are used to draw several broad conclusions regarding the implementation of environmental policy in the U. S. While the development and application of rules and regulations across the U.S. are examined in this chapter, special attention is paid to the evolution of policy addressing MTBE in California. The findings presented herein are based on research into the workings of selected state environmental management programs, and an in-depth assessment of recent developments in California, a state only now beginning to deal with the threat MTBE poses to drinking water supplies. Recent activity in California has helped trigger sweeping changes in national policy.

This chapter is divided into two sections. The first section presents an overview of policy-making approaches and analyzes the approaches taken by different states across the nation; the second takes a closer look at relevant events in California.

7.2 The making of environmental policy

Environmental policy can be made on the federal, state, or local level. On the federal level, policy is typically drafted by Congress and implemented by the U.S. EPA, or by state and local agencies empowered by the U.S. EPA to carry out these federally crafted environmental standards. State and local jurisdictions are usually granted license to append the federal mandates, though these agencies are typically not given the power to enforce a scheme that is less strict than that prescribed by the U.S. EPA. States and their authorized jurisdictions also establish action levels for chemicals that do not appear in the *Federal Register* — often referred to as "contaminants of concern."

Where recognized, MTBE has historically been regarded as a contaminant of concern, and as such has received myriad treatments by state and local governments charged with environmental protection. "Acceptable" concentrations of MTBE in groundwater vary widely across the U.S. and among regulatory agencies. The basis for these standards varies widely as well.

7.2.1 U.S. EPA guidelines for MTBE in groundwater

In December 1997, the U.S. EPA's Office of Water issued a drinking water advisory for MTBE. The advisory recommends a limit of 20 to 40 parts per billion (ppb) for drinking water as the range that is sufficiently low enough to prevent human health risk and reduce unpleasant taste and odor. The U.S. EPA has not yet established a maximum contaminant level (MCL) for MTBE.

The establishment of an MCL is a complex and lengthy process that, from start to finish, can take 10 years or more. The U.S. EPA's regulatory determination for a compound's MCL is based on (1) the human health risk of the contaminant, (2) the frequency with which the contaminant appears in drinking water supplies, and (3) the "meaningful" opportunity for health risk reductions achieved through regulation of the contaminant. Because of their timelines for research, review of available information about a contaminant of concern, regulatory determination, and rule development, the U.S. EPA's schedule to establish an MCL for MTBE was set for the year 2006 at the earliest.

In March 2000, the U.S. EPA issued a regulatory announcement stating its consideration of a limit or ban on the use of MTBE as a fuel additive, and published an Advance Notice of Proposed Rulemaking to issue a rule under Section 6 of the Toxic Substances Control Act by the end of 2000. As part of this process, the U.S. EPA will conduct an analysis of the risks and benefits of the use of MTBE, consider a range of options (including a total ban or a reduction in use), and issue their proposed rule addressing MTBE by the end of 2000. The U.S. EPA's final rule for MTBE will be issued after that. The schedule for the issuance of the final rule is difficult to predict as of the date of this publication.

7.3 Comparison of diverse state policies

The wide range of regulatory approaches to the hazards associated with the environmental release of MTBE can be seen in the results of research conducted by the authors of this publication in 1997 and 1998. The results of this survey of information available for different states' approaches to MTBE revealed three dominant trends in state policy relative to the management of MTBE releases at facilities storing and dispensing fuel from leaking underground fuel storage tanks (USTs). The first trend was seen in states that had established an advisory or cleanup level based on the U.S. EPA's health advisory and/or public reaction to water quality or air quality concerns. The second trend observed involved states with no clearly defined policy regulating MTBE, and often with no cleanup level for the compound, based on the perception that MTBE did not pose a significant risk to groundwater. The third group of states had arrived at an "acceptable" concentration of MTBE in groundwater, independent of the U.S. EPA's health studies and public complaint; most often for these states, the level at which MTBE was regulated hadn't changed much over a period of years.

See Appendix D, Summary of MTBE State-by-State Cleanup Standards, of this publication for a complete listing of state cleanup levels as of April 1, 2000.

7.3.1 First category: levels set according to U.S. EPA guidelines or public reaction

The first category — those states that had established a fairly rigorous cleanup or advisory level for MTBE based on the U.S. EPA guidelines or in response to public outcry — included the State of Florida. In 1997, Florida's action level for MTBE was changed to 35 ppb based on "general analytical criteria." According to the Florida Department of Health, the 1997 change was likely based on MTBE's possible cancer risk as determined by the U.S. EPA at the end of that year. MTBE cleanup levels for Florida were actually somewhat flexible according to site conditions and risk assessment, but usually fell within U.S. EPA's 20 to 40 ppb limit.

The maximum contaminant concentration for MTBE in groundwater was also changed during 1997 in New Jersey. New Jersey's MCL for MTBE was set at 700 ppb until February 1997, when it was decreased to 70 ppb. According to the New Jersey Division of Site Remediation — for which MTBE cleanup was something of a "sticking point" — the limit revision was based on new health studies, most of which had been released in 1996 and 1997. In response to the question of whether the widespread public condemnation of MTBE in New Jersey had affected groundwater policy, the Division stated that its policies were enacted in response to the results of new studies, not public perception.

7.3.2 Second category: no established cleanup level

Some states, such as Minnesota, Texas, and Oregon, did not have established cleanup levels for MTBE in groundwater at UST sites in 1997 and 1998. The results of the survey indicated that, for states in the second category, MTBE was not considered a sufficiently toxic or troublesome constituent to be labeled a "contaminant of concern."

In Minnesota, a state with a risk-based MTBE program, MTBE was included on a list of compounds to analyze during a groundwater contamination assessment. MTBE, however, was considered not so much a contaminant of concern as an early indicator of the need for cleanup, because MTBE tends to occur at the leading edge of a dissolved petroleum hydrocarbon plume. As such, MTBE was seen more as a diagnostic tool to locate a contaminant plume, as opposed to being considered a compound that triggers cleanup (unlike benzene, a gasoline component that is a U.S. EPA-listed carcinogen).

Like Minnesota, Oregon's MTBE program in 1997 was risk-based, and no established cleanup level existed for MTBE at sites where groundwater was contaminated with petroleum hydrocarbons. As research conducted in 1997 and 1998 revealed, the Oregon Department of Environmental Quality did not consider MTBE a contaminant of major concern. The Department referred requests for information about cleanup of MTBE to their website; however, cleanup of MTBE in groundwater was not mentioned at this website, although it did contain a description of Oregon's emphasis on risk-based assessment.

The Texas Department of Environmental Health likewise had no cleanup level for MTBE in groundwater, although a health advisory did exist. In Texas, unlike Minnesota, parties responsible for releases were not initially required to test for MTBE in groundwater, but were advised to treat the detection of MTBE in a drinking water supply well as an indication that cleanup of dissolved petroleum hydrocarbons was likely to be necessary.

Alaska is another example of a state that did not consider MTBE a serious risk to groundwater quality. Unlike the other states mentioned above, however, Alaska's policy was to not use MTBE in gasoline. The introduction of MTBE through Alaska's reformulated gas program commenced in 1995, lasting for only 2 months before public complaint — generally, reports of offensive odor, and respiratory and nervous system reactions — resulted in the state's elimination of MTBE from the reformulated gas program. The campaign to ban MTBE from use in Alaska was a local, state-specific effort. State advocacy groups and proponents of health studies collaborated with Alaska's Department of Health to ban MTBE. As of 1997, ethanol was Alaska's reformulated gas oxygenate of choice, and MTBE had not been detected in any of Alaska's groundwater supplies.

With the exception of Alaska, the water quality divisions of states in this category seemed to consider MTBE an important water quality issue only as an initial indicator of groundwater contamination by petroleum hydrocarbon constituents, not as a serious threat to groundwater quality and "consumer acceptability."

7.3.3 Third category: independently derived cleanup levels

The last category includes states with advisory and cleanup levels for MTBE that were derived independently from the U.S. EPA's health studies. Generally, these states had policies based on their own health and risk studies. The concentrations at which MTBE was regulated in groundwater in these states tended not to have changed much over a period of years.

The State of New York's guidance level, not so much a legislated cleanup level as a general reference for site closure criteria, was set at 50 ppb. This level had remained unchanged for nearly 10 years, and was based on established cleanup criteria for a "generic organic contaminant." Considerations of changes to this level, according to the New York State Health Department, would be based on studies independent of the U.S. EPA, coordinated by the health department, and approached on a case-by-case basis.

Wisconsin is another example of a state with a groundwater quality standard for MTBE (60 ppb) set by the state's health department according to a state-based toxicological formula. MTBE's threat to groundwater quality in

Table 7.1 Water Wells Impacted by MTBE in Santa Monica

System & Source Name	Sampling Date	MTBE Concentration (mg/L)
City of Santa Monica, Arcadia Well 05 (inactive)	Feb. 26, 1996 Apr. 22, 1996 May 28, 1996 June 24, 1996 July 22, 1996 Aug. 26, 1996	11.0 11.9 14.7 21.5 34.1 72.4
City of Santa Monica, Charnock Well 13	Feb. 26, 1996	130.0
City of Santa Monica, Charnock Well 18	May 28, 1996	30.6
City of Santa Monica, Charnock Well 19 (inactive)	Feb. 26, 1996	300.0

(Source: City of Santa Monica.)

part of Wisconsin, however, was mitigated because Wisconsin imposed a ban on MTBE in gasoline, effective only in Wisconsin's most populous urban area, Milwaukee.

7.3.4 Analysis of diverse approaches to MTBE policy

As noted above, in the absence of clear U.S. EPA directives, the approaches taken by states to the regulation of chemicals in the environment were varied. State agencies with no cleanup guidance for MTBE risk a reactive situation much like that seen in Santa Monica, California, where the sudden detection of MTBE in drinking water supply wells prompted public outcry and reactive agency response (see Table 7.1, above, for a listing of MTBE concentrations discovered in water wells in Santa Monica in 1997). Those states that have for a number of years considered MTBE a contaminant of concern, based on state-specific studies on health effects and fate and transport studies, seem to have fared the best. These states have formed their policies based on sound science under unified, rational agency oversight. These states' cleanup levels, while unaffected by the release of U.S. EPA's drinking water advisory, still tend to lie within the 20 to 40 ppb range specified by the advisory.

7.4 California: policy and transition

MTBE has been in use in California since 1979, but has recently come under intense scrutiny by legislators, the media, and a variety of regulatory agencies, both state and local. An examination of some of the recent events in

California pertaining to MTBE's impact on human health and the environment is useful when considering the development of environmental policy.

The positive detection of MTBE in the Santa Monica municipal water system in 1997 was discovered by accident. Reports surfacing in the months following the discovery suggested that the city's contracted water testing laboratory reported the presence of MTBE without being specifically directed to do so. According to these reports, the city usually analyzed the samples in-house, but during this sampling event used an outside lab; and it was an employee of this lab who noticed the MTBE spike on the chromatogram and asked the city if it desired quantification. No one knows how long the contaminant was in the city's water system prior to its discovery by the contract lab.

7.4.1 Regulatory framework

The limited scope of generally accepted California technical guidance documents on water quality may have been part of the problem the state faced in recognizing MTBE as a contaminant of concern. Even though MTBE was treated as a contaminant of concern by some other states (e.g., Florida) as far back as 1986, no provision was made for the monitoring of MTBE in California agency guidelines. The *California Leaking Underground Fuel Tank (LUFT) Manual* and regional companion documents (e.g., the Tri-Regional Appendix A to the LUFT, etc.) are silent on the need to test environmental samples for concentrations of MTBE. In fact, these rules and guidelines address only "total" hydrocarbon concentrations and volatile aromatics (benzene, etc.). These documents are further silent, with the exception of lead, on additives and lead scavengers (EDB, EDC) — indicating a somewhat unsophisticated approach to environmental monitoring. The LUFT manual was written in the mid-1980s, updated once, and has been on the drawing board for a second rewrite for the last 8 years. The State Water Resources Control Board (SWRCB), the lead agency regulating water quality in the state, finally issued draft MTBE management guidelines in December 1999, well after concerns over MTBE in the environment had become a public issue.

California's regulatory community has not historically acted in a unified manner, and communication between agencies, when it takes place, is not always effective. California overlooked the significance of MTBE and other additives because the SWRCB lacked a program of unified oversight, which has resulted in the technical isolation of many of the smaller, local regional boards and local rule-implementing agencies. These boards and agencies typically have the charter to make contacts outside their own jurisdiction, but don't always do so. The main technical resource for many of these isolated agencies has been the outdated LUFT manual.

Given the historic disunity exhibited by regulatory agencies in California, it is not difficult to understand the impediments to the evolution of policy addressing MTBE. The mid- to late-1990s were marked by a sudden uproar over the additive after the Santa Monica discovery, ultimately resulting in drastic modification to state environmental policy. What actually prompted the Santa Monica discovery and the ensuing reaction to MTBE?

7.4.2 Grassroots political action leads to state legislative action

In the mid-1990s, groups of mostly UST owners led by the vocal Santa Rosa-based Environmental Resources Council began voicing their opposition to what they believed to be overly punitive UST-related environmental regulations. In response to these concerns, Senator Michael Thompson of Santa Rosa authored Senate Bill (SB) 1764, which was signed into law by Governor Pete Wilson early in 1995. SB 1764 required the SWRCB to "convene an advisory committee consisting of distinguished chemists, biologists, health professionals, geologists, engineers, and other appropriate professionals" to advise the SWRCB on issues pertaining to the state's UST program. A separate UST program evaluation effort was also initiated by the SWRCB, and involved the commissioning of a team of individuals led by the Lawrence Livermore National Laboratory (LLNL). The report produced by the LLNL team, "California Leaking Underground Fuel Tank (LUFT) Historical Case Analyses," released in 1995, was pushed to the fore by the SWRCB. The work of the SB 1764 committee, which included recommendations regarding MTBE, vanished into obscurity.

7.4.3 Release of LLNL report casts spotlight on MTBE

The LLNL report was the first of two supported by SWRCB-financed initiatives. The report received a great deal of positive publicity, owing primarily to its main conclusion that UST cleanup efforts, beyond contamination source removal, were largely unproductive and, by association, unnccessary. In fact, the SWRCB went so far as to release the LLNL report with a memorandum from Walt Petit, Executive Director, proclaiming that the hazard posed by gasoline releases into the environment was not nearly as severe as it was initially thought to be. Thanks largely to the LLNL report and the "Petit Memo," regional water boards and local agencies put many cleanup enforcement actions for UST sites on hold, and began to evaluate all gasoline contamination events as potential "low-risk" cases. As might be expected, the LLNL report, the Petit Memo, and the ensuing changes in agency policy (albeit "interim") were met with acclaim from industry groups, and protests from environmental advocates.

The lists of positive and negative LLNL study attributes were loudly heralded by interested parties on both sides of the issue. One "poster child" identified by the anti-LLNL faction was MTBE, due to the above-mentioned fact that the LLNL study was absolutely silent on the subject of gasoline additives. This observation quickly gained notoriety as those opposed to the LLNL findings worked to discredit the report.

The identification of this deficiency in the LLNL report was the first time MTBE was paid serious public attention in California. Ironically, the LLNL report, a document that seemed to dismiss the hazards associated with gasoline leaks and spills, by its omission of any discussion of MTBE, accomplished exactly the opposite.

As the MTBE debate grew, LLNL and the SWRCB conceded that MTBE and other additives had been ignored due to the lack of agency requirements for their detection and monitoring in the environment. As asserted later by agencies in the state, no specific data existed pertaining to the fate, transport, or human health effects of additives, so they were not included in the LLNL study.

The connection between the work by the SWRCB to evaluate the California UST program and the onset of MTBE-related publicity seems apparent. That an issue such as the use and regulation of a gasoline additive rises to public attention in California through such a process is ironic, particularly since MTBE and similar compounds have been treated as the equivalent of agency-controlled substances by other states for almost a decade. California's regulatory community has, in the past, been a national leader in areas of environmental regulation. Nonetheless, as described by a regional water board staff member in an address to a San Francisco Bay Area professional organization in 1998, the state was clearly "asleep at the switch" when it came to MTBE.

7.4.4 California policy issued: closing the barn door

Public reaction upon learning of California agencies' oversight in addressing MTBE made headlines, and it quickly became apparent that the California regulatory community would need to respond aggressively. Respond they did, with a raft of proposed legislation and interim agency guidelines — though the responses were generally silent in regard to the weaknesses in the bureaucratic structure that led to the problem in the first place.

In July 1996, the SWRCB issued a memorandum to the California regional boards and local oversight UST program managers. The SWRCB memo requested that MTBE be added to the list of gasoline components in contaminated groundwater being monitored at gasoline UST release sites. In 1997, California issued their guideline cleanup level of 35 ppb. California instituted this new guideline based partly on pressures from a public fearful of the real or imagined consequences of MTBE, rather than on science and reasoned policy-making.

As could be expected, California politicians also responded to the accounts of the dangers posed by MTBE with anti-MTBE legislation. Several bills relating to the study and regulation of MTBE were passed by the California state legislature and signed into law. These bills include SB 521, authored by Senator Mountjoy, which appropriated $500,000 to the University of California for "a specified study and assessment of the human health and environmental risks and benefits, if any, of MTBE, to be submitted to the governor by January 1, 1999." Another bill, SB 1189 (Hayden), proposed to begin the state's process of the adoption of primary and secondary drinking water standards for MTBE on or before July 1, 1998, as well as the study for a possible state listing of MTBE as a carcinogenic or reproductive toxin.

AB 592 (Kuehl) also provided for the establishment of a subaccount within the state's Underground Storage Tank Cleanup Fund dedicated to reimbursing public water systems for the cost of treatment of the water supply or the provision of alternate drinking water supplies. AB 2439 (Bowen) proposed the prevention of the discharge of gasoline from watercraft propelled by a two-stroke engine in a lake or reservoir that serves as a domestic water supply.

A federal bill, H.R. 3518 (Bilbray), introduced in May 1996, is pending in the U.S. House of Representatives as of the date of this publication. This bill would effectively repeal the federally mandated requirement for California to use reformulated gasoline, if toxics and ozone emissions can be reduced without applying RFG requirements.

7.4.5 Epilogue

In March 2000, the U.S. EPA issued their recommendation for a phase-out and ultimate ban of MTBE. In doing this, the U.S. EPA has effectively recognized that, in order to satisfy the 1990 CAA, substitutes for MTBE (such as ethanol) will likely have to be developed and marketed. The U.S. EPA has also asked Congress to work with them in efforts to reduce or eliminate the use of MTBE in the U.S., possibly by modifying or suspending the oxygenate requirements mandated by the 1990 CAA. With this action by the U.S. EPA, it appears as if future MTBE-related federal policy activity will address the mitigation of damage already done through the use of MTBE in gasoline.

The analysis of environmental policy as it has been applied to MTBE in California and the U.S. raises many questions about the overall structure, source, and application of environmental policy in other arenas, such as:

- Is it possible to thoroughly evaluate the potential health and environmental impacts of a substance such as MTBE prior to its introduction and use?
- How often do well-intended legislative efforts (such as the 1990 CAA) pressure industry to implement hasty and poorly conceived compliance measures? How often will these attempts to comply create a bigger problem than the one originally addressed?
- Recognizing the importance of states' rights, would it be prudent for the U.S. EPA or another federal authority to develop a mechanism to educate and monitor agencies implementing environmental standards across the U.S. to at least ensure a consistent, minimum standard of environmental regulation?

As the headlines fade, political attention turns away; purveyors of news media seem convinced that Americans want to read about calamity and drama, and stories reporting (and thereby motivating) positive pre-emptive reform rarely make it to the front page. If the national policy construct remains as it was through the MTBE crisis — and if policy on a state level continues to be reactive, inconsistent, and based on outdated information or unsound science — a repeat environmental policy performance is almost certain to occur.

Endnotes and references

[1] Zogorski, J.S. et al., Fuel Oxygenates and Water Quality: Current Understanding of Sources, Occurrence in Natural Water, Environmental Behavior, Fate and Significance, Interagency Oxygenated Fuel Assessment, Office of Science and Technology, Washington, D.C., 1996.
[2] Sakata, R., A Drinking Water Standard for MTBE???, U.S. EPA, Office of Water, 1999.

chapter eight

Conclusions and recommendations

8.1 Future groundwater quality deterioration from use of MTBE

Approximately half of the U.S. domestic water supply comes from groundwater. The remaining half is provided by lakes, rivers, and reservoirs. Until this point, the national water supply has been of a very high caliber owing to the existence of plentiful, high quality resources and effective pollution prevention efforts by state and federal agencies. Ensuring the quality of this resource has been the goal of federal, state, and local regulators. The introduction of MTBE into the national water supply represents a significant threat to that quality because of MTBE's persistence in the environment and because of the potential for MTBE to accumulate in the nation's aquifers.

Concerns related to MTBE highlight other important groundwater issues including inadequate underground storage tank standards, variable tank installation quality, inadequate wellhead protection and abandonment standards, materials compatibility issues, and incomplete combustion of fuels in engines. These concerns must ultimately be addressed if the nation is to maintain the high quality of its groundwater resources.

The problem of MTBE contamination in groundwater will not be eliminated overnight by a statewide or even a federal ban on this gasoline additive, and must be dealt with by water resource professionals. At the very least, the economic impact of the contamination with MTBE of future drinking water supplies will likely enhance the financial prospects for the bottled water industry and owners of water rights. Figure 8.1 shows the frequency of detection of MTBE in shallow groundwater from a 1993 to 1994 USGS study.

Water resource management practices and the national attitude toward water resources are changing, and likely will continue to change owing, in part, to concerns over MTBE contamination. Mixing, or dilution of, MTBE-contaminated water with clean water to reach acceptably low levels of con-

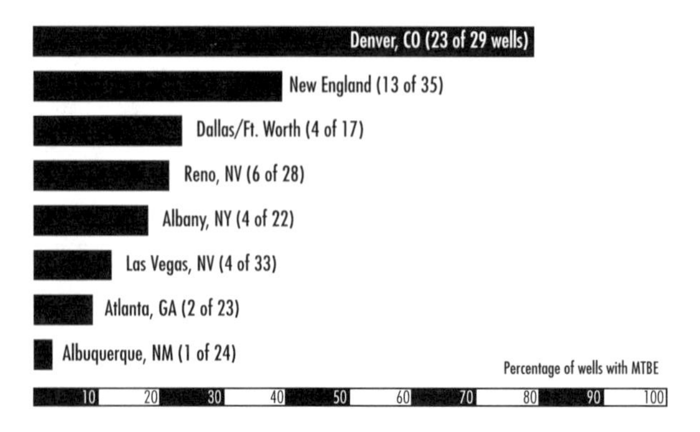

Figure 8.1 Frequency of detection of MTBE in shallow groundwater, 1993 to 1994. (Source: USGS Open-File Report 95-456, 1995.)

tamination is occurring in some isolated areas in the nation; this practice is likely to become more commonplace over the next few years as municipalities grapple with MTBE issues. If the use of MTBE is not reduced or stopped in a timely manner, and if groundwater and surface water degradation increases from MTBE accumulation in drinking-water supplies, recycled (or "gray") water may become a significant water source. The potential for use of recycled water will require enormous public education efforts and a major shift in consumer attitudes toward recycled water. Nonetheless, the use of recycled water may become more attractive to municipalities trying to meet nonpotable local water demands, in order to free up potable water supplies for drinking.

If MTBE significantly degrades water quality, the implementation of technologies — such as desalinization, advanced oxidation processes, or point of use treatment systems — will be required to maintain the nation's existing high quality drinking water supplies. These expensive processes are highly dependent on the skillful marketing of advanced or nascent technologies, the effective implementation of advances in water provider operations, and the control of energy costs.

Ironically, MTBE was initially used in gasoline as a technology solution for an air pollution problem. After the introduction and during the use of MTBE, national air quality standards have improved; yet, at the same time, regional groundwater resources have been affected by MTBE, and these resources, regionally as well as on the whole, are arguably at risk for serious degradation.

The uneasy balance between the privilege of personal freedom afforded by the automobile and the preservation of our natural resources has been debated ad infinitum since the introduction of the automobile. A new twist has been introduced into the discussion with the detection of MTBE in groundwater resources, causing the nation to once again re-evaluate the trade-offs between gasoline usage and environmental quality. Ultimately, we must either accept advanced technologies to clean up MTBE-affected groundwater, pay a higher cost for better quality storage tanks that won't leak, or pay significantly higher costs for water of guaranteed quality from bottled water vendors or owners of large-scale water resources.

8.2 Conclusions

The following are some conclusions that may be formed after an evaluation of the problem of MTBE in groundwater.

- Gasoline additives, including MTBE, ethanol, and other butene derivatives, have had an undetermined effect in reducing vehicle emissions.
- While MTBE does present some human health risk, the risk has not been well-defined owing to a lack of conclusive studies regarding the human health risk of MTBE. It is possible that the actual cancer and noncancer health risks are low. Moreover, the concentration threshold for the turpentine-like taste and unpleasant odor of MTBE is lower than any health advisory threshold, and, therefore, ingestion of MTBE at concentrations greater than these thresholds is unlikely.
- MTBE is highly soluble in water, and is chemically stable in the environment. In addition, MTBE travels at virtually the same velocity as groundwater, faster than the less mobile BTEX compounds. Thus, remediation of groundwater resources contaminated by MTBE will be costly and difficult.
- Leaks and spills of gasoline containing MTBE from a variety of point sources, including tanker trucks, pipelines, underground tanks, and above-ground tanks will continue into the future, regardless of legislation and engineering controls to prevent such occurrences. The current, more stringent tank regulations, effective December 22, 1998, should continue to be effective in reducing future releases of MTBE into subsurface soils and groundwater.
- The recent phase-out of MTBE introduced by California Governor Gray Davis has already begun to reduce MTBE's harmful potential in that state, but the use of MTBE will not be completely phased out until 2002. Until then, MTBE will continue to have the potential to pose a threat to groundwater resources.
- The national concerns about MTBE highlight other important groundwater issues, including inadequate underground storage tank standards, variable tank installation quality, inadequate wellhead protection and abandonment standards, materials incompatibility issues,

and incomplete combustion of fuels in engines. These concerns must ultimately be addressed if the nation is to protect its high quality groundwater resources from MTBE contamination as well as contamination by other substances.

8.3 Challenges for the future

The following challenges have been highlighted by the discovery of MTBE in surface and groundwater, along with potential solutions:

- Improperly constructed or abandoned wells, or poorly sealed wells, can become conduits in the subsurface for MTBE or other chemicals. Adequate wellhead protection standards and well abandonment programs that require high quality seals for all wells must be implemented nationwide.
- The compatibility of MTBE with the various compounds used for connections, seals, joints, pipes, and coatings in hoses, engines, and underground storage tank systems has not yet been resolved. Without such a resolution, the potential for MTBE to contaminate groundwater may not be eliminated, even after tank upgrades are completed.
- Incomplete combustion and poor engine performance, most notably in currently designed 2-cycle engine, allows leakage of fuel products, including MTBE, into surface waters such as lakes and rivers. Redesign of these types of engines, or a reduction in their use, may reduce leakage.
- Ethanol, another oxygenating fuel additive, will likely replace MTBE in gasoline in states such as California. Further studies of the potential for the use of ethanol in gasoline, including health, air, and water quality studies, should be conducted.

8.4 Recommendations

Although the health issues associated with MTBE have been fiercely debated, it is the opinion of the authors of this publication that health concerns are of secondary importance to the potential degradation of beneficial uses of the nation's potable aquifers by releases of gasoline containing MTBE. Once even very low concentrations of MTBE are detected in a potable aquifer, that groundwater resource can be determined to be significantly degraded based on the actual or potential breach of taste and odor thresholds; therefore, its use becomes restricted. Because up to 50% of the nation's drinking water comes from groundwater sources, it is prudent to ban the use of a chemical that can destroy the usefulness of groundwater to this degree.

Based on the difficult and costly nature of remediating MTBE in groundwater, and the unknown cumulative amount of MTBE in the environment, the writers of this report recommend that the use of MTBE be phased out nationally. Life-cycle analyses of MTBE and other oxygenate alternatives should be performed to include combustion efficiency and storage considerations, as well as the transport and fate of the various substances in air, soil, groundwater, and surface water environments. The results from the life-cycle analyses should be used to select the best additive, if one is needed for future use.

MTBE will persist in the environment for several decades to come. The solutions to air quality problems caused by vehicle emissions should include consideration of not replacing MTBE with another oxygenate. Other remedies for vehicle air pollution include the establishment of cleaner-burning engines and cleaner-burning fuels with fewer impurities. Never again should a water resources disaster of national magnitude be created through blindly embracing a technology intended to solve air quality problems.

Appendix A

Glossary of technical terms and acronyms used in this book

Acute Exposure — A short-term exposure to a chemical, usually consisting of a single exposure or dose administered for a period of less than 24 hours (acute : chronic as short-term : long-term).
Adsorb — To gather (a gas, liquid, or dissolved substance) on a surface in a condensed layer, as when charcoal adsorbs gases.
Advection — The usually horizontal movement of a mass of fluid (air, for example); transport by such movement.
Animal Cancer Bioassay — A long-term experimental study in which animals are given high doses of a chemical in order to obtain information on the carcinogenicity of the chemical in test animals. The likelihood that a chemical will cause cancer in humans at the low doses typical of human exposures is estimated from information obtained from animal cancer bioassays.
Annular Space/Annulus — The space in a soil boring or monitoring well between the casing and the borehole wall.
Benzene — A volatile toxic liquid aromatic hydrocarbon used as a solvent and in gasoline. Benzene has been determined to cause cancer in humans exposed to high concentrations.
CPF — Cancer potency factor (also called "slope" factor) in units of $(mg/[kg\ body\ mass]/d)^{-1}$.
Cal EPA — California Environmental Protection Agency
Carbon Monoxide (CO) — A colorless, very toxic gas that is formed as a product of the incomplete combustion of carbon and carbon-based fuels such as gasoline.
Carcinogen — A substance or agent producing or inciting cancer.
Carcinogenicity — The ability of a chemical to induce cancer by reacting with genetic material or by interfering with other normal biological processes of the body's cells.
Chronic Exposure — A long-term exposure to a chemical for a period of 1 year or more in animals, and more than 7 years in humans.
Clean Air Act — Legislation originally passed by the U.S. Congress in 1976, and amended in 1990. Among other provisions, the 1990 Amendments mandate the use of oxygenated fuels and reformulated gasoline in nonattainment areas (areas with poor air quality). The Clean Air Act led to the establishment

of air quality standards, including air quality criteria that are based on what is known about the effects of air pollutants on animals, humans, plants, and materials.

Co-Elute — Refers to substances whose retention times in a chromatographic column are the same. The substances exit the column at the same time; therefore, they co-elute, and the detection of one substance through chromatographic analysis may be masked by another.

Diffusion — The gradual mixing of the molecules of two or more substances owing to random molecular motion, as in the dispersion of a vapor in air.

Dose — The quantity of a substance administered to the body over a specified time period — units: mg/(kg body mass)/d or mg/[(kg body mass)d].

Dose-response Assessment — A step in the assessment of risks to humans from potentially toxic agents, in which the relationship between the dose levels to which animals or humans are exposed and the health-effect responses at each dose level are characterized in a quantitative manner.

Elute — To remove (adsorbed material) from an adsorbent by means of a solvent.

Epidemiology — A branch of medical science that deals with the incidence, distribution, and control of disease in a population; the sum of the factors controlling the presence or absence of a disease or pathogen.

Exposure — Contact of an organism with one or more substances.

Exposure route — Mode of substance bodily intake: inhalation, dermal sorption, ingestion.

Exposure pathway — Mode of substance transport from source to receptor through air, soil, water, and the food chain.

Formaldehyde — A reaction product in the photooxidation of hydrocarbons and a primary metabolite, or breakdown product, of MTBE. Formaldehyde is classified as a primary irritant at high-dose environmental exposures.

Gavage — Introduction of material into the stomach by a tube.

Incremental Risk — Risk from a specific cause over and above the total risk from all causes.

Inhalation Exposure — Intake of a chemical or substance (e.g., dust) into the body and the lungs through breathing.

Intake rate — Quantity of a substance taken into the body per unit time — units: mg/(kg body mass)/d or mg/[(kg body mass)d].

LC_{50} — Lethal concentration that is estimated to kill 50% of control laboratory animals from a specified inhalation exposure duration (usually 96 hours).

LD_{50} — Lethal dose that is estimated to kill 50% of control laboratory animals from a specified ingestion exposure duration (usually 96 hours).

LOAEL — Lowest observed adverse effect level.

MCL — Maximum contaminant level (for contaminant in water).

MRL — Minimal risk level.

MTBE — Methyl *tertiary*-butyl ether is an organic liquid containing oxygen that is added to reformulated gasoline and fuel in order to reduce the emissions of toxics into the air. MTBE is the most widely used oxygenate in the U.S.

Metabolism — The sum of the processes by which a particular substance is handled in the living body.
Metabolite — A product of metabolism; the compound formed when a parent compound is metabolized as a result of biological processes in the body.
NOAEL — No observed adverse effect level.
Oxygenated Fuel — Gasoline to which additives containing oxygen have been added. Oxygenated fuel differs from reformulated gasoline in the amount of oxygenate (2.7% by mass, 15% by volume) that it contains. (See Reformulated Gasoline.)
Ozone (O_3) — A molecule consisting of three oxygen atoms that is a major air pollutant in the lower atmosphere. Ozone is a primary component of urban smog, resulting from the combustion of gasoline.
Part per million — A way of expressing low concentrations of a substance in a medium such as air or water; for example, 1 part per million MTBE means that there is 1 unit of MTBE in a million units of air, soil, or water.
Part per billion — A way of expressing very low concentrations of a substance in a medium such as air or water; for example, 1 part per billion MTBE means that there is 1 unit MTBE in a billion units of air, soil, or water.
Pharmacokinetics — The study of the bodily absorption, distribution, metabolism, and excretion of a drug or chemical.
Proposition 65 — Safe Drinking Water & Toxic Enforcement Act of 1986 (California).
Reference dose (RfD)/Reference concentration (RfC) (inhalation) — The U.S. EPA-approved intake rate for a substance that is unlikely to result in significant adverse noncancer (mostly acute but also chronic) effects.
Reformulated gasoline — Gasoline to which additives containing oxygen have been added. The amended Clean Air Act requires the year-round use of reformulated gasoline in areas with unhealthful levels of ozone. Reformulated gasoline differs from oxygenated fuel in the amount of oxygenate (2.0% by mass, 11% by volume) that it contains. (See Oxygenated Fuel.)
Risk — The statistical probability that a particular adverse effect will occur, measured in dimensions of frequency or incidence (i.e., one in one million).
Risk assessment — The assessment of the likelihood and potential magnitude of adverse health effects associated with human exposures to agents such as toxic chemicals or radiation. The four steps of risk assessment are: hazard identification, toxicity assessment, exposure assessment, and risk characterization.
***Tertiary*-butyl alcohol** — *Tertiary*-butyl alcohol (TBA) is a primary metabolite of MTBE in the body. TBA may be associated with some adverse health effects observed in animals.
Threshold dose — The dose above which a chemical or other toxic agent produces an adverse health effect and below which no adverse health effects are seen or anticipated.
Toxicity —The ability of an agent, such as a chemical or drug, to cause harmful health effects to living organisms.

Toxicology — A science that deals with poisons and their effects and with the problems (clinical, industrial, or legal, for example) involved therein.
Vadose Zone — A term used in geology indicating a soil zone found or located above the water table. (The term "unsaturated zone" also fits this definition.)

Sources

Merriam-Webster's Collegiate Dictionary, www.m-w.com/cgi-bin/dictionary, 2000
Jacques Guertin
Christy Herron

Appendix B

Conversions for international system (SI metric) and United States units

Area

Square Millimeters	Square centimeters (cm^2)	0.01
	Square inches (in^2)	1.55×10^3
Square Centimeters	Square millimeters (mm^2)	100.0
	Square meters (m^2)	$1. \times 10^{-4}$
	Square inches	0.1550
	Square feet (ft^2)	1.07639×10^{-3}
Square Inches	Square millimeters	645.16
	Square centimeters	6.4516
	Square meters	6.4516×10^{-4}
	Square feet	69.444×10^{-4}
Square feet	Square meters	0.0929
	Hectares (ha)	9.2903×10^{-6}
	Square inches	144.0
	Acres	2.29568×10^{-5}
Square yards	Square meters	0.83613
	Hectares	8.3613×10^{-5}
	Square feet	9.0
	Acres	2.0612×10^{-4}
Square meters	Hectares	1.0×10^{-4}
	Square feet	10.76391
	Acres	2.47×10^{-4}
	Square yards (yd^2)	1.19599
Acres	Square meters	4046.8564
	Hectares	0.40469
	Square feet	4.356×10^4
Hectares	Square meters	1.0×10^4
	Acres	2.471

Square kilometers	Square meters	1.0×10^6
	Hectares	100.0
	Square feet	107.6391×10^5
	Acres	247.10538
	Square miles (Mi^2)	0.3861
Square miles	Meters	258.998
	Hectares	81×10^4
	Square kilometers (km^2)	258.99881
	Square feet	2.58999
	Acres	2.78784×10^7

Force per unit area, pressure-stress

Pounds per square inch	Kilopascals (kPa)	6.89476
	Meters-head [(a)]	0.70309
	Mm of Hg [(b)]	51.7151
	Feet of water [(b)]	2.3067
	Pounds per square foot (lb/ft^2)	144.0
	Std. atmospheres	68.046×103
Pounds per square foot	Kilopascals	0.04788
	Meters-head [(a)]	4.8826×10^{-3}
	Mm of Hg [(b)]	0.35913
	Feet of water [(a)]	16.0189×10^{-3}
	Pounds per square inch	6.9444×10^{-3}
	Std. Atmospheres	0.47254×10^{-3}
Short tons per square foot	Kilopascals	95.76052
	Pounds per square inch (lb/in^2)	13.88889
Meters-head [(a)]	Kilopascals	9.80636
	Mm of Hg [(b)]	73.554
	Feet of water [(a)]	3.28084
	Pounds per square inch	1.42229
	Pounds per square foot	204.81
Feet of water [(a)]	Kilopascals	2.98898
	Meters-head [(a)]	0.3048
	Mm of Hg [(b)]	22.4193
	Inches of Hg [(b)]	0.88265
	Pounds per square inch	0.43351
	Pounds per square foot	62.4261
Kilopascals	Newtons per square meter (N/m^2)	1.0×10^3
	Mm of Hg [(b)]	7.50064
	Meters-head [(a)]	0.10197
	Inches of Hg [(b)]	0.2953
	Pounds per square foot	20.8854
	Pounds per square inch	0.14504
	Std. atmospheres	9.8692×10^{-3}

Kilograms (f) per square meter	Kilopascals	$9.80665 - 10^{-3}$
	Mm of Hg [b]	73.556×10^{-3}
	Pounds per square inch	1.4223×10^{-3}
Millibars (mbars)	Kilopascals	0.10
Bars	Kilopascals	100.0
Std. Atmospheres	Kilopascals	101.325
	Mm of Hg [b]	760.0
	Pounds per square inch	14.70
	Feet of water	33.90

Length

Micrometers	Millimeters	1.0×1^{-3}
	Meters	1.0×1^{-6}
	Angstrom units (Å)	1.0×10^{-4}
	Mils	.03937
	Inches	3.93701×10^{-5}
Millimeters	Micrometers	1.0×10^{3}
	Centimeters (cm)	0.1
	Meters	1.0×10^{-3}
	Mils	39.37008
	Inches	.03937
	Feet (ft)	3.28084×10^{-3}
Centimeters	Millimeters	10.0
	Meters	0.01
	Mils	0.3937×10^{3}
	Inches	0.3937
	Feet	0.03281
Inches	Millimeters	25.40
	Meters	0.0254
	Mils	1.0×10^{3}
	Feet	0.08333
Feet	Millimeters	304.8
	Meters	0.3048
	Inches	12.0
	Yards (yd)	0.33333
Yards	Meters	0.9144
	Inches	36.0
	Feet	3.0
Meters	Millimeters	1.0×10^{-3}
	Kilometers (km)	1.0×10^{-3}
	Inches	39.37008
	Yards	1.09361
	Miles (mi)	6.21371×10^{-4}

Kilometers	Meters	1.0×10^3
	Feet	3.28084×10^3
	Miles	0.62137
Miles	Meters	1.60934×10^3
	Kilometers	1.60934
	Feet	5280.0

Mass

Grams	Kilograms (kg)	1.0×10^{-3}
	Ounces (avdp)	0.03527
Ounces (avdp)	Grams (g)	28.34952
	Kilograms	0.02835
	Pounds (avdp)	0.0625
Pounds (avdp)	Kilograms	0.45359
	Ounces (avdp)	16.00
Kilograms	Kilograms (force)-second squared per meter (kgf.s²/m)	0.10197
	Pounds	2.20462
	Slugs	0.06852
Slugs	Kilograms	14.5939
Short tons	Kilograms	907.1847
	Metric tons (t)	0.90718
	Pounds (avdp)	2000.0
	Yards	1760.0
Metric tons (tonne or megagram)	Kilograms	1.0×10^3
	Pounds (avdp)	2.20462×10^3
	Short tons	1.10231

Mass per unit volume, density, and mass capacity

Pounds per cubic foot	Kilogram per cubic meter (kg/m³)	16.01846
	Slugs per cubic foot (slug/ft³)	0.03108
	Pounds per gallon (lb/gal)	0.13368
Pounds per gallon	Kilograms per cubic meter (kg/m³)	119.8264
	Slugs per cubic foot	0.2325
Pounds per cubic yard	Kilograms per cubic meter	0.59328
	Pounds per cubic foot (lb/ft³)	0.03704
Grams per cubic centimeter	Kilograms per cubic meter	1.0×10^3
Ounces per gallon (oz/gal)	Grams per liter (g/l)	7.48915
Kilograms per cubic meter	Grams per cubic centimeter (g/cm³)	1.0×10^{-3}
	Pounds per cubic foot (lb/ft³)	1.0×10^{-3}
	Pounds per gallon	62.4297×10^{-3}
	Pounds per cubic yard	1.68556

Long tons per cubic yard	Kilograms per cubic meter	1328.939
Ounces per cubic inch (oz/in^3)	Kilograms per cubic meter	1729.994
Slugs per cubic foot	Kilograms per cubic meter	515.3788

Velocity

Feet per second	Centimeter per square meter (cm^2)	3.11×10^{-4}
	Square feet (ft^2)	3.35×10^{-7}
	Meters per second (m/s)	0.3048
	Kilometers per hour (km/h)	1.09728
	Miles per hour (mi/h)	0.68182
Meters per second	Centimeters per square meter (cm^2)	1.02×10^{-3}
	Square feet (ft^2)	1.10×10^{-6}
	Feet per second (ft/s)	3.28
	Kilometers per hour	3.60
	Feet per second (ft/s)	3.28084
	Miles per hour	2.23694
Centimeters per squared meter (cm^2)	Square feet (ft^2)	1.08×10^{-3}
	Meters per second (m/s)	9.80×10^{2}
	Feet per second (ft/sec)	3.22×10^{3}
Square feet (ft^2)	Centimeters per squared meter (cm^2)	9.29×10^{2}
	Meters per second (m/s)	9.11×10^{5}
	Feet per second (ft/sec)	2.99×10^{6}
Kilometers per hour	Meters per second	0.27778
	Feet per second	0.91134
	Miles per hour	0.62147
Miles per hour	Kilometers per hour	1.690934
	Meters per second	0.44704
	Feet per second	1.46667

Viscosity

Centipoise	Pascal-second (Pas)	1.0×10^{-3}
	Poise	0.01
	Pound per foot-hour (lb/ft * h)	2.41909
	Pound per foot-section (lb/ft *s)	6.71969×10^{-4}
	Slug per foot-second (slug/ft * s)	2.08854×10^{-5}
Pascal-second	Centipoise	1000.0
	Pound per foot-hour	2.41990×10^{9}
	Pound per foot-second	0.67197
	Slug per foot-second	20.885×10^{-3}
Pounds per foot-hour	Pascal-second	4.13379×10^{-4}
	Pound per foot-second	2.77778×10^{-4}

Pounds per foot-second	Pascal-second	1.48816
	Slug per foot-second	31.0809×10^{-3}
	Centipoise	1.48816×10^{3}
Centistokes	Square meters per second (m^2/s)	1.0×10^{-6}
	Square feet per second (ft^2/s)	10.76391×10^{-6}
	Stokes	0.01
Square feet per second	Square meters per second	9.2903×10^{-2}
	Centistokes	0.2903×10^{4}
Stokes	Square meters per second	1.0×10^{-4}
Rhe	1 per pascal-second (1/Pas)	10.0

Volume-capacity

Cubic millimeters	Cubic centimeters (cm^3)	1.0×10^{-3}
	Liters (l)	1.0×10^{-6}
	Cubic inches (in^3)	61.02374×10^{-6}
Cubic centimeters	Liters	1.0×10^{-6}
	Milliliters (ml)	1.0
	Cubic inches (in^3)	61.02374×10^{-6}
	Fluid ounces (fl.oz)	33.814×10^{-6}
Milliliters	Liters	1.0×10^{-6}
	Cubic centimeters	1.0
Cubic inches	Milliliters	16.38706
	Cubic feet (ft^3)	57.87037×10^{-5}
Liters	Cubic meters	1.0×10^{-3}
	Cubic feet	0.03531
	Gallons	0.26417
	Fluid ounces	33.814
Gallons	Liters	3.78541
	Cubic meters	3.78541×10^{-3}
	Fluid ounces	128.0
	Cubic feet	0.13368
Cubic feet	Liters	28.31685
	Cubic meters (m^3)	28.31685×10^{-3}
	Cubic dekameters (dam^3)	28.31685×10^{-6}
	Cubic inches	1728.0
	Cubic yards (yd^3)	37.03704×10^{-3}
	Gallons (gal)	7.48052
	Acre-feet (acre-ft)	22.95684×10^{-6}
Cubic miles	Cubic dekameters	4.16818×10^{6}
	Cubic kilometers (km^3)	4.16818
	Acre-feet	3.3792×10^{6}
Cubic yards	Cubic meters	0.76455
	Cubic feet	27.0

Cubic meters	Liters	1.0×10^3
	Cubic dekameters	1.0×10^{-3}
	Gallons	264.1721
	Cubic feet	35.31467
	Cubic yards	1.30795
	Acre-feet	8.107×10^{-4}
Acre-feet	Cubic meters	1233.482
	Cubic dekameters	1.23348
	Cubic feet	43.560×10^3
	Gallons	325.8514×10^3
Cubic dekameters	Cubic meters	1.0×10^3
	Cubic feet	35.31467×10^3
	Acre-feet	0.81071
	Gallons	26.41721×10^4
Cubic kilometers	Cubic dekameters	1.0×10^6
	Acre-feet	0.8107×10^6
	Cubic miles (mi^3)	0.23991

Volume per cross-sectional area per unit time: transmissivity [(a)]

Cubic feet per foot per day ($ft^3/(ft*d)$)	Cubic meters per meter per day ($m^3/(m*d)$)	0.0929
	Gallons per foot per day ($gal/(ft*d)$)	7.48052
	Liters per meter per day ($1/(m*d)$)	92.903

Gallons per foot per day per day ($m^3/(m*d)$)	Cubic meters per meter	0.01242
	Cubic feet per foot per day ($ft^3/(ft*d)$)	0.13368

Volume per unit area per unit time: hydraulic conductivitiy (permeability)

Cubic feet per square foot per day	Cubic meters per square meter per day (m^3/m^2d)	0.3048
	Cubic feet per square foot per minute ($ft^3/ft^2/min$)	0.6944×10^{-3}
	Liters per square meter per day ($l/m^2/d$)	304.8
	Gallons per square foot per day ($gal/ft^2/d$)	7.48052
	Cubic millimeters per square millimeter per day ($mm^3/mm^2/d$)	304.8
	Cubic millimeters per square millimeter per hour ($mm^3/mm^2/h$)	25.4
	Cubic inches per square inch per hour ($in^3/in^2/h$)	0.5

Gallons per square foot per day	Centimeter per square meter (cm^2)	5.42×10^{-10}
	Cubic meters per square meter per day ($m^3/m^2/d$)	40.7458×10^{-3}
	Liters per square meter per day ($l/(m^2/d)$)	40.7458
	Cubic feet per square foot per day ($ft^3/ft^2/d$))	0.13368

Volume per unit time flow

Cubic feet per second	Liters per second (l/s)	28.31685
	Cubic meters per second (m^3/s)	0.02832
	Cubic dekameters per day (dam^3/d)	2.44657
	Gallons per minute (gal/min)	448.83117
	Acre-feet per day (acre-ft/d)	1.98347
	Cubic feet per minute (ft^3/min)	60.0
Gallons per minute	Cubic meters per second	0.631×10^{-4}
	Liters per second	0.0631
	Cubic dekameters per day	5.451×10^{-3}
	Cubic feet per second (ft^3/s)	2.228×10^{-3}
	Acre-feet per day	4.4192×10^{-3}
Acre-feet per day	Cubic meters per second	0.01428
	Cubic dekameters per day	1.23348
	Cubic feet per second	0.50417
Cubic dekameters per day	Cubic meters per second	0.01157
	Cubic feet per second	0.40874
	Acre-feet per day	0.81071

(a) Column of H^2O (water) measured at 4°C.
(b) Column of Hg (mercury) measured at 0°C.
(c) Many of these units can be dimensionally simplified: for example, $m^3/(m*d)$ can also be written m^3/d).

Appendix C

Material safety data sheets: MTBE and gasoline

Material safety data sheet: MTBE

Date Printed: 04/27/00

Dates Valid: 05/2000 to 07/2000

Section 1 – Product and Company Information

Product Name	TERT-BUTYL METHYL ETHER, 99.8%, HPLC GRADE
Brand	Aldrich Chemical
Company	Aldrich Chemical Co., Inc.
Street Address	1001 West St. Paul
City, State, Zip, Country	Milwaukee, WI 53233 USA
Telephone	414-273-3850

Section 2 – Chemical Identification

CATALOG #: 293210

NAME: TERT-BUTYL METHYL ETHER, 99.8%, HPLC GRADE

Section 3 – Composition/Information on Ingredients

CAS #: 1634-04-4

MF: C5H12O

EC NO: 216-653-1

Synonyms

TERT-BUTYL METHYL ETHER * 2-METHOXY-2-METHYLPROPANE * METHYL TERT-BUTYL ETHER (ACGIH) * METHYL 1,1-DIMETHYLETHYL ETHER * PROPANE, 2-METHOXY-2-METHYL- (9CI) *

Section 4 – Hazards Identification

LABEL PRECAUTIONARY STATEMENTS

FLAMMABLE (USA)

HIGHLY FLAMMABLE (EU)

HARMFUL

IRRITATING TO EYES, RESPIRATORY SYSTEM AND SKIN.

POSSIBLE RISK OF IRREVERSIBLE EFFECTS.

TARGET ORGAN(S):

KIDNEYS

NERVES

KEEP AWAY FROM SOURCES OF IGNITION - NO SMOKING.

IN CASE OF CONTACT WITH EYES, RINSE IMMEDIATELY WITH PLENTY OF

WATER AND SEEK MEDICAL ADVICE.

WEAR SUITABLE PROTECTIVE CLOTHING.

HYGROSCOPIC

HANDLE AND STORE UNDER NITROGEN.

REFRIGERATE BEFORE OPENING

Section 5 – First-Aid Measures

IN CASE OF CONTACT, IMMEDIATELY FLUSH EYES OR SKIN WITH COPIOUS AMOUNTS OF WATER FOR AT LEAST 15 MINUTES WHILE REMOVING CONTAMINATED CLOTHING AND SHOES. IF INHALED, REMOVE TO FRESH AIR. IF NOT BREATHING GIVE ARTIFICIAL RESPIRATION. IF BREATHING IS DIFFICULT, GIVE OXYGEN. IF SWALLOWED, WASH OUT MOUTH WITH WATER PROVIDED PERSON IS CONSCIOUS. CALL A PHYSICIAN. WASH CONTAMINATED CLOTHING BEFORE REUSE

Section 6 – Fire Fighting Measures

EXTINGUISHING MEDIA: CARBON DIOXIDE, DRY CHEMICAL POWDER OR APPROPRIATE FOAM. WATER MAY BE EFFECTIVE FOR COOLING, BUT MAY NOT EFFECT EXTINGUISHMENT.

SPECIAL FIREFIGHTING PROCEDURES: WEAR SELF-CONTAINED BREATHING APPARATUS AND PROTECTIVE CLOTHING TO PREVENT CONTACT WITH SKIN AND EYES. USE WATER TRAY TO COOL FIRE-EXPOSED CONTAINERS.

FLAMMABLE LIQUID.

UNUSUAL FIRE AND EXPLOSIONS HAZARDS

VAPOR MAY TRAVEL CONSIDERABLE DISTANCE TO SOURCE OF IGNITION AND FLASH BACK.

CONTAINER EXPLOSION MAY OCCUR UNDER FIRE CONDITIONS.

FORMS EXPLOSIVE MIXTURES IN AIR.

Section 7 – Accidental Release Measures

EVACUATE AREA.

SHUT OFF ALL SOURCES OF IGNITION.

WEAR SELF-CONTAINED BREATHING APPARATUS, RUBBER BOOTS AND HEAVY RUBBER GLOVES.

COVER WITH AN ACTIVATED CARBON ADSORBENT, TAKE UP AND PLACE IN CLOSED CONTAINERS. TRANSPORT OUTDOORS.

VENTILATE AREA AND WASH SPILL SITE AFTER MATERIAL PICKUP IS COMPLETE.

Section 8 – Handling and Storage

REFER TO SECTION 7.

Section 9 – Exposure Controls/Personal Protection

WEAR APPROPRIATE NIOSH/MSHA-APPROVED RESPIRATOR, CHEMICAL-RESISTANT GLOVES, SAFETY GOGGLES, OTHER PROTECTIVE CLOTHING.

MECHANICAL EXHAUST REQUIRED.

SAFETY SHOWER AND EYE BATH.

USE NONSPARKING TOOLS.

DO NOT BREATHE VAPOR.

DO NOT GET IN EYES, ON SKIN, ON CLOTHING.

AVOID PROLONGED OR REPEATED EXPOSURE.

WASH THOROUGHLY AFTER HANDLING.

KEEP TIGHTLY CLOSED.

KEEP AWAY FROM HEAT, SPARKS, AND OPEN FLAME.

STORE IN A COOL DRY PLACE.

Section 10 – Physical and Chemical Properties

PHYSICAL PROPERTIES

BOILING POINT: 55 C TO 56 C

FLASHPOINT –27 F
–2 C

EXPLOSION LIMITS IN AIR:

UPPER 15.1%

LOWER 1.6%

AUTOIGNITION TEMPERATURE: 705 F 373 C

VAPOR PRESSURE: 4.05PSI 20 C 14.78PSI 55 C

SPECIFIC GRAVITY: 0.740

VAPOR DENSITY: 3.1

Section 11 – Stability and Reactivity

STABILITY: STABLE.

INCOMPATIBILITIES: OXIDIZING AGENTS, STRONG ACIDS, PROTECT FROM MOISTURE.

HAZARDOUS COMBUSTION OR DECOMPOSITION PRODUCTS

TOXIC FUMES OF: CARBON MONOXIDE, CARBON DIOXIDE

HAZARDOUS POLYMERIZATION WILL NOT OCCUR.

Section 12 – Toxicological Information

ACUTE EFFECTS: HARMFUL IF SWALLOWED, INHALED, OR ABSORBED THROUGH SKIN. VAPOR OR MIST IS IRRITATING TO THE EYES, MUCOUS MEMBRANES AND UPPER RESPIRATORY TRACT. CAUSES SKIN IRRITATION.

EXPOSURE CAN CAUSE: NAUSEA, VOMITING, DIZZINESS, CNS DEPRESSION. ASPIRATION OR INHALATION MAY CAUSE CHEMICAL PNEUMONITIS. RAPIDLY ABSORBED FOLLOWING ORAL EXPOSURE.

CHRONIC EFFECTS: THIS PRODUCT IS OR CONTAINS A COMPONENT THAT HAS BEEN REPORTED TO BE POSSIBLY CARCINOGENIC BASED ON ITS IARC, ACGIH, NTP OR EPA CLASSIFICATION.

TARGET ORGAN(S): KIDNEYS, CENTRAL NERVOUS SYSTEM. TO THE BEST OF OUR KNOWLEDGE, THE CHEMICAL, PHYSICAL, AND TOXICOLOGICAL PROPERTIES HAVE NOT BEEN THOROUGHLY INVESTIGATED.

ADDITIONAL INFORMATION

MTBE (METHYL-TERT-BUTYL ETHER) IS REPORTED TO METABOLIZE TO TERT-BUTYL ALCOHOL AND FORMALDEHYDE BY MICROSOMAL DEMETHYLATION. ACCORDING TO AN ARTICLE IN TOXICOLOGY AND INDUSTRIAL HEALTH, VOLUME II, NUMBER 2, PAGES 119-149, 1995; MTBE (METHYL-TERT-BUTYL ETHER) SHOULD BE CONSIDERED A "POTENTIAL HUMAN CARCINOGEN" DUE TO "AN INCREASE IN LEYDIG INTERSTITIAL CELL TUMORS OF TESTES IN MALE RATS" AND AN "INCREASE IN LYMPHOMAS, LEUKEMIAS, AND UTERINE SARCOMAS" IN FEMALE RATS.

IN ANOTHER UNPUBLISHED STUDY MTBE WAS SHOWN TO BE CARCINOGENIC DUE TO "INCREASED INCIDENCE OF A RARE TYPE OF KIDNEY TUMOR" IN MALE RATS AND AN "INCREASE IN THE INCIDENCE OF HEPATOCELLULAR ADENOMAS" IN FEMALE MICE.

RTECS #: KN5250000

ETHER, TERT-BUTYL METHYL

TOXICITY DATA

ORL-RAT LD50:4 GM/KG NTIS** PB87-174603

IHL-RAT LC50:23576 PPM/4H NTIS** PB87-174603

ORL-MUS LD50:5960 UL/KG CHHTAT 70,172,1990

IHL-MUS LC50:141 GM/M3/15M ANESAV 11,455,1950

IPR-MUS LD50:1700 UL/KG CHHTAT 70,172,1990

ONLY SELECTED REGISTRY OF TOXIC EFFECTS OF CHEMICAL SUBSTANCES (RTECS) DATA IS PRESENTED HERE. SEE ACTUAL ENTRY IN RTECS FOR COMPLETE INFORMATION.

Section 13 – Ecological Information

DATA NOT YET AVAILABLE.

Section 14 – Disposal Considerations

BURN IN A CHEMICAL INCINERATOR EQUIPPED WITH AN AFTERBURNER AND SCRUBBER BUT EXERT EXTRA CARE IN IGNITING AS THIS MATERIAL IS HIGHLY FLAMMABLE. OBSERVE ALL FEDERAL, STATE AND LOCAL ENVIRONMENTAL REGULATIONS.

Section 15 – Transport Information

CONTACT ALDRICH CHEMICAL COMPANY FOR TRANSPORTATION INFORMATION.

Section 16 – Regulatory Information

EUROPEAN INFORMATION

HIGHLY FLAMMABLE

HARMFUL

R 1

HIGHLY FLAMMABLE

R 36/37/38

IRRITATING TO EYES, RESPIRATORY SYSTEM AND SKIN.

R 40

POSSIBLE RISK OF IRREVERSIBLE EFFECTS.

S 16

KEEP AWAY FROM SOURCES OF IGNITION - NO SMOKING.

S 26

IN CASE OF CONTACT WITH EYES, RINSE IMMEDIATELY WITH PLENTY OF WATER AND SEEK MEDICAL ADVICE.

S 36

WEAR SUITABLE PROTECTIVE CLOTHING.

REVIEWS, STANDARDS, AND REGULATIONS

OEL=MAK

ACGIH TLV-CONFIRMED ANIMAL CARCINOGEN DTLVS* TLV/BEI, 1999

ACGIH TLV-TWA 40 PPM DTLVS* TLV/BEI, 1999

OEL-RUSSIA:STEL 100 MG/M3 JAN 1993

OEL-SWEDEN: TWA 50 PPM (180 MG/M3), STEL 75 PPM (250 MG/M3), JAN 1999

NOES 1983: HZD X4267; NIS 6; TNF 614; NOS 9; TNE 5996; TFE 1783

EPA TSCA SECTION 8(B) CHEMICAL INVENTORY

EPA TSCA SECTION 8(D) UNPUBLISHED HEALTH/SAFETY STUDIES ON EPA IRIS DATABASE

EPA TSCA TEST SUBMISSION (TSCATS) DATA BASE, DECEMBER 1999

NIOSH ANALYTICAL METHOD, 1994: METHYL TERT-BUTYL ETHER, 1615

U.S. INFORMATION

THIS PRODUCT IS SUBJECT TO SARA SECTION 313 REPORTING REQUIREMENTS.

Section 17 – Other Information

THE ABOVE INFORMATION IS BELIEVED TO BE CORRECT BUT DOES NOT PURPORT TO BE ALL INCLUSIVE AND SHALL BE USED ONLY AS A GUIDE. SIGMA, ALDRICH, FLUKA SHALL NOT BE HELD LIABLE FOR ANY DAMAGE RESULTING FROM HANDLING OR FROM CONTACT WITH THE ABOVE PRODUCT. SEE REVERSE SIDE OF INVOICE OR PACKING SLIP FOR ADDITIONAL TERMS AND CONDITIONS OF SALE.

COPYRIGHT 1999 SIGMA-ALDRICH CO.

LICENSE GRANTED TO MAKE UNLIMITED PAPER COPIES FOR INTERNAL USE ONLY

(From Sigma-Aldrich Co., 1999. With permission.)

Material safety data sheet: gasoline range organics

Date Printed: 04/27/00

Dates Valid: 10/1999 - 12/1999

Section 1 – Product and Company Information

Product Name UST MODIFIED GRO 1X1ML MEOH 1000UG/ML EACH

Company	Supelco Inc.
Street Address	Supelco Park
City, State, Zip, Country	Bellefonte, PA 16823-0048 USA
Telephone	814-359-3441

Section 2 – General Information

CATALOG NO 48167 (REORDER PRODUCT BY THIS NO.)

DATA SHEET NO I481670

UST MODIFIED GASOLINE RANGE ORGANICS

FORMULA MIXTURE FORMULA WEIGHT

CAS NRTECS

SYNONYM ANALYTICAL STANDARD IN METHANOL

MANUFACTURER SUPELCO INC. PHONE 814-359-3441

ADDRESS SUPELCO PARK, BELLEFONTE, PA 16823-0048

Section 3 – Hazardous Ingredients of Mixtures

CHEMICAL NAME

COMMON NAME - PERCENTAGE - CAS #

(FORMULA) - PEL(UNITS) - TLV(UNITS)

LD50 VALUE - CONDITIONS

BENZENE, 1,3,5-TRIMETHYL-

1,3,5-TRIMETHYLBENZENE 0.1 108-67-8

C6H3-1,3,5-(CH3)3 N/A 25 PPM

NAPHTHALENE

NAPHTHALENE 0.1 91-20-3

C10H8 10 PPM 10 PPM

490 MG/KG ORAL RAT SEE FOOTNOTE(6)

BENZENE, ETHYL-

ETHYLBENZENE 0.1 100-41-4

C2H5C6H5 100 PPM 100 PPM

3500 MG/KG ORAL RAT SEE FOOTNOTE(6)

0-XYLENE

O-XYLENE 0.1 95-47-6

CH3C6H4CH3 100 PPM 100 PPM

5000 MG/KG ORAL RAT SEE FOOTNOTE(6)

BENZENE, 1,3-DIMETHYL-

M-XYLENE 0.1 108-38-3

CH3C6H4CH3 100 PPM 100 PPM

5000 MG/KG ORAL RAT SEE FOOTNOTE(6)

BENZENE, 1,4-DIMETHYL-

P-XYLENE 0.1 106-42-3

CH3C6H4CH3 100 PPM 100 PPM

5000 MG/KG ORAL RAT SEE FOOTNOTE(6)

PROPANE, 2-METHOXY-2-METHYL-

METHYL TERT-BUTYL ETHER 0.1 1634-04-4

C5H12O N/A 50 PPM

4000 MG/KG ORAL RAT SEE FOOTNOTE(6)

BENZENE, 1,2,4-TRIMETHYL

1,2,4-TRIMETHYLBENZENE 0.1 95-63-6

C9H12 N/A N/A

6000 MG/KG ORAL RAT SEE FOOTNOTE(6)

METHANOL

METHANOL 99.0 67-56-1

CH3OH 260 MG/M3 262 MG/M3

5628 MG/KG ORAL RAT SEE FOOTNOTE(6)

BENZENE

BENZENE 0.1 71-43-2

C6H6 1 PPM 10 PPM

4894 MG/KG ORAL RAT SEE FOOTNOTE(1,5,6,7)

BENZENE, METHYL-

TOLUENE 0.1 108-88-3

C6H5CH3 100 PPM 50 PPM

5000 MG/KG ORAL RAT SEE FOOTNOTE(6)

FOOTNOTES

[1]CLASSIFIED BY IARC AS A CLASS 1 CARCINOGEN.

[5]OSHA REGULATED CARCINOGEN, 29 CFR 1910.

[6]SUBJECT TO THE REPORTING REQUIREMENTS OF SARA TITLE III, SECTION 313.

[7]CLASSIFIED BY NTP AS A GROUP A CARCINOGEN.

Section 4 – Physical Data

BOILING POINT 65 C MELTING POINT -98 C

VAPOR PRESSURE 100 MM VAPOR DENSITY (AIR=1) 1.10

SPECIFIC GRAVITY .790 G/ML C (WATER=1) PERCENT VOLATILE BY VOLUME 100

WATER SOLUBILITY 100 EVAPORATION RATE >1 (ETHER=1)

APPEARANCE: CLEAR COLORLESS LIQUID

Section 5 – Fire and Explosion Hazard Data

FLASH POINT 50 F FLAMMABLE LIMITS LEL 6.0 UEL 36.5

EXTINGUISHING MEDIA:

C02

DRY CHEMICAL

ALCOHOL FOAM.

SPECIAL FIRE FIGHTING PROCEDURES:

WEAR SELF CONTAINED BREATHING APPARATUS WHEN FIGHTING A CHEMICAL FIRE.

UNUSUAL FIRE AND EXPLOSION HAZARDS:

THE FOLLOWING TOXIC VAPORS ARE FORMED WHEN THIS MATERIAL IS HEATED TO DECOMPOSITION.

CARBON MONOXIDE, FORMALDEHYDE.

Section 6 – Health Hazard Data

LD50 5628 MG/KG ORAL RAT TLV 262 MG/M3

PEL 260 MG/M3

EMERGENCY AND FIRST AID PROCEDURES:

EYES — FLUSH EYES WITH WATER FOR 15 MINUTES. CONTACT A PHYSICIAN.

SKIN — FLUSH SKIN WITH LARGE VOLUMES OF WATER. WASH CLOTHING AND SHOES BEFORE REUSING.

INHALATION — IMMEDIATELY MOVE TO FRESH AIR. GIVE OXYGEN IF BREATHING IS LABORED. IF BREATHING STOPS, GIVE ARTIFICIAL RESPIRATION. CONTACT A PHYSICIAN.

INGESTION — NEVER GIVE ANYTHING BY MOUTH TO AN UNCONSCIOUS PERSON. NEVER TRY TO MAKE AN UNCONSCIOUS PERSON VOMIT. GIVE 2 TABLESPOONS OF BAKING SODA IN A GLASS OF WATER. PRESS FINGERS TO BACK OF TONGUE TO INDUCE VOMITING.

IMMEDIATELY CONTACT A PHYSICIAN.

EFFECTS OF OVEREXPOSURE — HARMFUL IF INHALED. MAY BE FATAL IF SWALLOWED. CONTAINS MATERIAL(S) KNOWN TO THE STATE OF CALIFORNIA TO CAUSE CANCER.

LOW BLOOD PRESSURE (HYPOTENSION).

DERMATITIS

BREATHING DIFFICULTY

HEADACHE

NAUSEA

DIZZINESS

GASTROINTESTINAL DISTURBANCES

DEPRESSES CENTRAL NERVOUS SYSTEM

NARCOSIS

LIVER DAMAGE

KIDNEY DAMAGE

RESPIRATORY FAILURE

BLINDNESS

LEUKEMIA

MAY CAUSE CORNEAL INJURY.

Section 7 – Reactivity Data

STABILITY: STABLE.

CONDITIONS TO AVOID: N/A

INCOMPATIBILITY: STRONG ACIDS

OXIDIZING AGENTS: CHROMIC ANHYDRIDE, LEAD PERCHLORATE, PERCHLORIC ACIDS, ALUMINUM AND MAGNESIUM, METALS

HAZARDOUS DECOMPOSITION PRODUCTS: CARBON MONOXIDE, FORMALDEHYDE.

HAZARDOUS POLYMERIZATION WILL NOT OCCUR.

CONDITIONS TO AVOID: N/A

Section 8 – Spill or Leak Procedures

STEPS TO BE TAKEN IN CASE MATERIAL IS RELEASED OR SPILLED: TAKE UP WITH ABSORBENT MATERIAL. VENTILATE AREA. ELIMINATE ALL IGNITION SOURCES.

WASTE DISPOSAL METHOD: COMPLY WITH ALL APPLICABLE FEDERAL, STATE, OR LOCAL REGULATIONS

Section 9 – Special Protection Information

RESPIRATORY PROTECTION (SPECIFIC TYPE): WEAR FACE MASK WITH ORGANIC VAPOR CANISTER.

PROTECTIVE GLOVES: WEAR RUBBER GLOVES.

EYE PROTECTION: WEAR PROTECTIVE GLASSES.

VENTILATION: USE ONLY IN WELL VENTILATED AREA.

SPECIAL: N/A

OTHER PROTECTIVE EQUIPMENT: N/A

Section 10 – Special Precautions

STORAGE AND HANDLING: REFRIGERATE IN SEALED CONTAINER. KEEP AWAY FROM OXIDIZERS. KEEP AWAY FROM IGNITION SOURCES.

OTHER PRECAUTIONS: AVOID EYE OR SKIN CONTACT. AVOID BREATHING VAPORS. WHILE THE INFORMATION AND RECOMMENDATIONS SET FORTH HEREIN ARE BELIEVED TO BE ACCURATE AS OF THE DATE HEREOF, SUPELCO, INC. MAKES NO WARRANTY WITH RESPECT THERETO AND DISCLAIMS ALL LIABILITY FROM RELIANCE THEREON.

LAST REVISED 1/01/99

COPYRIGHT 1999 SUPELCO, INC. SUPELCO PARK BELLEFONTE, PA 16823-0048 8864

ALL RIGHTS RESERVED

(With permission.)

Appendix D

Summary of MTBE state-by-state cleanup standards

April 1, 2000

State	**Cleanup level**
Alabama	20 mg/L
Alaska	EPA*
Arizona	35 µg/L; EPA*
Arkansas	Site Specific
California	13 µg/L primary (health effects) 5 µg/L secondary (taste and odor)
Colorado	Site Specific
Connecticut	100 µg/L
Delaware	180 µg/L
Florida	50 µg/L residential, 500 µg/L industrial
Georgia	EPA*
Hawaii	20 µg/L
Idaho	Site Specific
Illinois	70 µg/L projected
Indiana	45 µg/L projected
Iowa	EPA*
Kansas	20 µg/L
Kentucky	EPA*
Lousiana	18 µg/L
Maine	35 µg/L
Maryland	Site Specific; 20 µg/L for drinking water
Massachusetts	70 µg/L groundwater drinking supply 50,000 µg/L groundwater source for vapor emissions to buildings

Michigan	40 µg/L projected
Minnesota	40 µg/L; EPA
Mississippi	55 µg/L; EPA
Missouri	40 µg/L to 400 µg/L (depending on use)
Montana	30 µg/L
Nebraska	EPA*
Nevada	20 µg/L; 200 µg/L
New Hampshire	70 µg/L
New Jersey	70 µg/L
New Mexico	100 µg/L
New York	10 µg/L
North Carolina	200 µg/L; proposed at 70
North Dakota	EPA*
Ohio	40 µg/L
Oklahoma	20 µg/L
Oregon	20 µg/L; 40 µg/L
Pennsylvania	20 µg/L
Rhode Island	40 µg/L groundwater quality standard; 20 µg/L preventative action level
South Carolina	40 µg/L
South Dakota	EPA*
Tennessee	EPA*
Texas	EPA*
Utah	200 µg/L action level
Vermont	40 µg/L
Virginia	Site Specific
Washington	20 µg/L
West Virginia	EPA*
Wisconsin	60 µg/L
Wyoming	200 µg/L

NOTES:

EPA* Waiting for EPA to set cleanup levels

Cleanup levels can change at any time.

Appendix E

Geologic principles and MTBE

E.1 Introduction

Successful subsurface characterization, detection monitoring, and ultimate remediation of methyl *tertiary*-butyl ether (MTBE) are predicated on a solid conceptual understanding of geology and hydrogeology. The factors affecting fate and transportation of MTBE, determination of possible adverse risks to public health, safety, and welfare or the degradation of groundwater resources are largely controlled by regional and local subsurface conditions. Successful resolution of MTBE contamination requires adequate geologic and hydrogeologic characterization leading to insights of the preferential migration pathways followed by a development of an appropriate remediation strategy.

Geologic factors control the vertical and lateral movement, distribution, and quality of groundwater as well as MTBE movement through physical, chemical, and biological processes. Fate and transport of MTBE in soils and aquifers are, to a large part, controlled by the lithology, stratigraphy, and structure of the geologic deposits and formations. The rate of migration of MTBE in the subsurface is dependent on numerous factors.

Modeling of MTBE migration through preferred pathways relies on both regional and local scale evaluations and understanding. Regional scale might describe a fault-bounded structural basin containing dozens of vertically and laterally adjacent depositional facies. Each depositional facies might be composed of rocks containing some of the same building blocks: gravels, sands, silts, and clays, in different configurations and juxtapositions.

Facies distribution and changes in distribution are dependent on a number of interrelated controls: sedimentary process, sediment supply, climate, tectonics, sea level changes, biological activity, water chemistry, and volcanism. The relative importance of these regional factors ranges between different depositional environments (Reading, 1978). However, on the local and sub-local scale, the porosity and permeability of a particular sandstone aquifer in a specific site in that basin might depend on factors such as individual grain size, grain size sorting, and primary and secondary cementation.

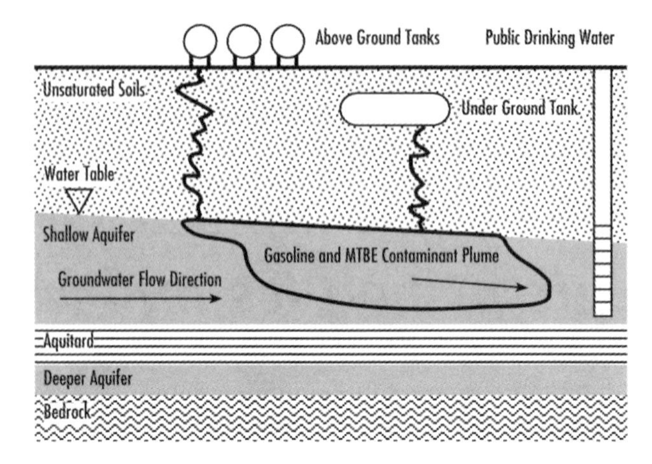

Figure E.1 Interaction of geologic control and contaminant plume migration. (From Geoprobe Systems, 1999. With permission.)

Lithology and stratigraphy are the most important factors affecting MTBE movement in soils and unconsolidated sediments. Stratigraphic features including geometry and age relations between lenses, beds and formations, and lithologic characteristics of sedimentary rocks such as physical composition, grain size, grain packing, and cementation are among the most important factors affecting groundwater and MTBE flow in sedimentary rocks. Igneous-metamorphic rocks are geologic systems produced by deformation after deposition or crystallization. Groundwater and MTBE flow in igneous-metamorphic rocks is most affected by structural features such as cleavages, fractures, folds, and faults (Freeze and Cherry, 1979). Figure E.1 shows a conceptual representation of the transport of petroleum constituents or MTBE from the surface through various layers of the subsurface.

The largest percentage of environmentally contaminated sites in the world lie on alluvial and coastal plains consisting of complex interstratified sediments. The majority of these contaminated sites have some component of impacted shallow soils or unconsolidated sediments. Shallow groundwater contamination by MTBE can result from surface (aboveground tank or tanker leakage) or near-surface activities (underground storage tank leakage). Deeper soils or rocks may also become impacted owing to preferred flow pathways along fault zones, a lack of a competent aquitard to stop migrating MTBE, or unintended manmade conduits such as abandoned mines or improperly designed or abandoned wells.

Data collected from subsurface investigations can be compiled to obtain a three-dimensional framework to use in developing a corrective action plan.

Contamination potential maps can be developed based on several parameters including

- Depth to shallow aquifers (i.e., 50 ft or less).
- Hydrogeologic properties of materials between the aquifers and ground surface.
- Relative potential for geologic material to transmit water.
- Description of surface materials and sediments.
- Soil infiltration data.
- Presence of deeper aquifers.
- Potential for hydraulic intercommunication between aquifers.

Development of other types of geologic, hydrogeologic, and MTBE concentration maps can be used for the preliminary screening of sites for the storage, treatment, or disposal of hazardous and toxic materials. Such maps focus on those parameters that are evaluated as part of the screening process. Maps exhibit outcrop distribution of rock types that may be suitable as host rocks; distribution of unconsolidated, water-bearing deposits; distribution and hydrologic character of bedrock aquifers; and regional recharge/discharge areas. These maps can thus be used to show areas where special attention needs to be given for overall waste management including permitting of new facilities, screening of potential new disposal sites or waste management practices, and the need for increased monitoring of existing sites and activities in environmentally sensitive areas.

Overall understanding of the regional geologic and hydrogeologic framework, characterization of regional geologic structures, and proper delineation of the relationship between various aquifers is essential to implementing both short- and long-term aquifer restoration and rehabilitation programs, and assessing aquifer vulnerability. Understanding of the regional hydrogeologic setting can serve in designation of aquifers for beneficial use, determination of the level of remediation warranted, implementation of regional and local remediation strategies, prioritization of limited manpower and financial resources, and overall future management. Preferred fluid migration pathways are influenced by porosity and permeability, sedimentary sequences, facies architecture, and fractures.

E.2 Hydrogeology

Henry Darcy performed the first studies of groundwater flow in 1856. The French engineer was most interested in one-dimensional flow of water through vertically oriented pipes filled with sand. What he determined is now known as Darcy's law: the flow is proportional to the cross-sectional area and the head loss along the pipe and inversely proportional to the flow length. In addition, Darcy determined that the quantity of flow is proportional to the

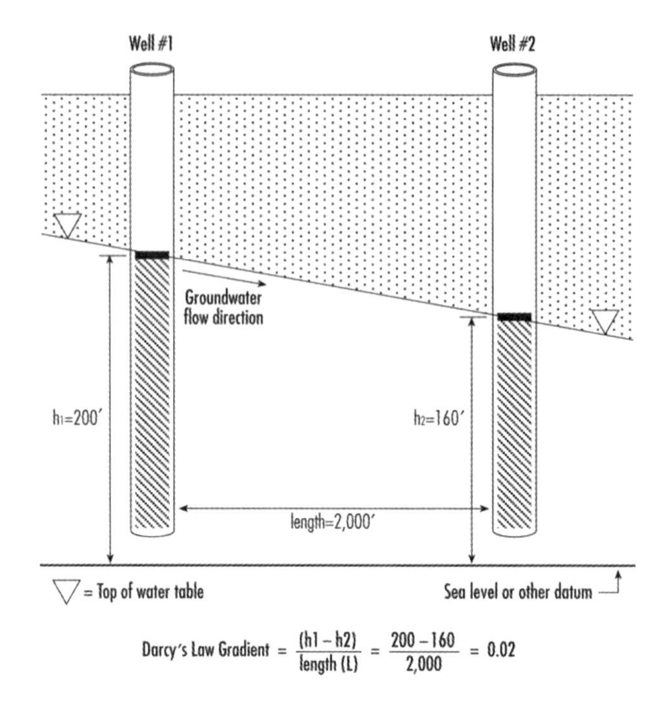

Figure E.2 Groundwater gradient calculated using Darcy's Law.

coefficient (K), which is also related to the nature of the porous material. Darcy's Law states:

$$Q = -KA\ dh/dl$$

where

- Q is the volumetric discharge or flow rate in gallons per day (gpd).
- K is the hydraulic conductivity in gpd/ft^2.
- A is the cross-sectional area in ft^2.
- dh/dl or i is the hydraulic gradient, which does not have units.

Another way to write Darcy's Law is $Q = -KiA$. Groundwater gradients are calculated using Darcy's Law as shown in Figure E.2.

E.3 Porosity, permeability, and diagenesis

Groundwater occurs in aquifers, which can consist of sedimentary deposits such as sands or gravels, or lithified units such as sandstones or conglomerates. Water is stored in the pore spaces in the sediments and rocks. Porosity is the measure of the percent of a material that is void space.

More specifically, porosity is a measure of pore space per unit volume of rock or sediment and can be divided into two types: absolute porosity and effective porosity. Absolute porosity (n) is the total void space per unit volume and is defined as the percentage of the bulk volume that is not solid material. The equation for basic porosity is listed below

$$n = \frac{\text{bulk volume} - \text{solid volume}}{\text{bulk volume}} \times 100 \qquad \text{(eq. 1)}$$

Porosity can be individual open spaces between sand grains in a sediment or fracture spaces in a dense rock. A fracture in a rock or solid material is an opening or a crack within the material. Matrix refers to the dominant constituent of the soil, sediment, or rock, and is usually a finer-sized material surrounding or filling the interstices between larger-sized material or features. Gravel may be composed of large cobbles in a matrix of sand. Likewise, a volcanic rock may have large crystals floating in a matrix of glass. The matrix will usually have different properties than the other features in the material. Often, either the matrix or the other features will dominate the behavior of the material, leading to the terms matrix-controlled transport, or fracture-controlled flow.

Effective porosity (N_e) is defined as the percentage of the interconnected bulk volume (i.e., void space through which flow can occur) that is not solid material. The equation for effective porosity is listed below:

$$N_e = \frac{\text{interconnected pore volume}}{\text{bulk volume}} \times 100 \qquad \text{(eq. 2)}$$

Effective porosity (N_e) is of more importance and, along with permeability (the ability of a material to transmit fluids), determines the overall ability of the material to readily store and transmit fluids or vapors. Where porosity is a basic feature of sediments, permeability is dependent upon the effective porosity, the shape and size of the pores, pore interconnectiveness (throats), and properties of the fluid or vapor. Fluid properties include capillary force, viscosity, and pressure gradient (see Figure E.3).
Specific yield is closely related to effective porosity.

$$Sy = \frac{\text{Vw drained}}{\text{Vt}}$$

Specific yield (Sy) is the ratio of the volume of water (Vw) drained from saturated soil or rock due to the attraction of gravity equal to the effective porosity divided by the total volume (Vt).

$$Sr = \frac{\text{Vw retained}}{\text{Vt}}$$

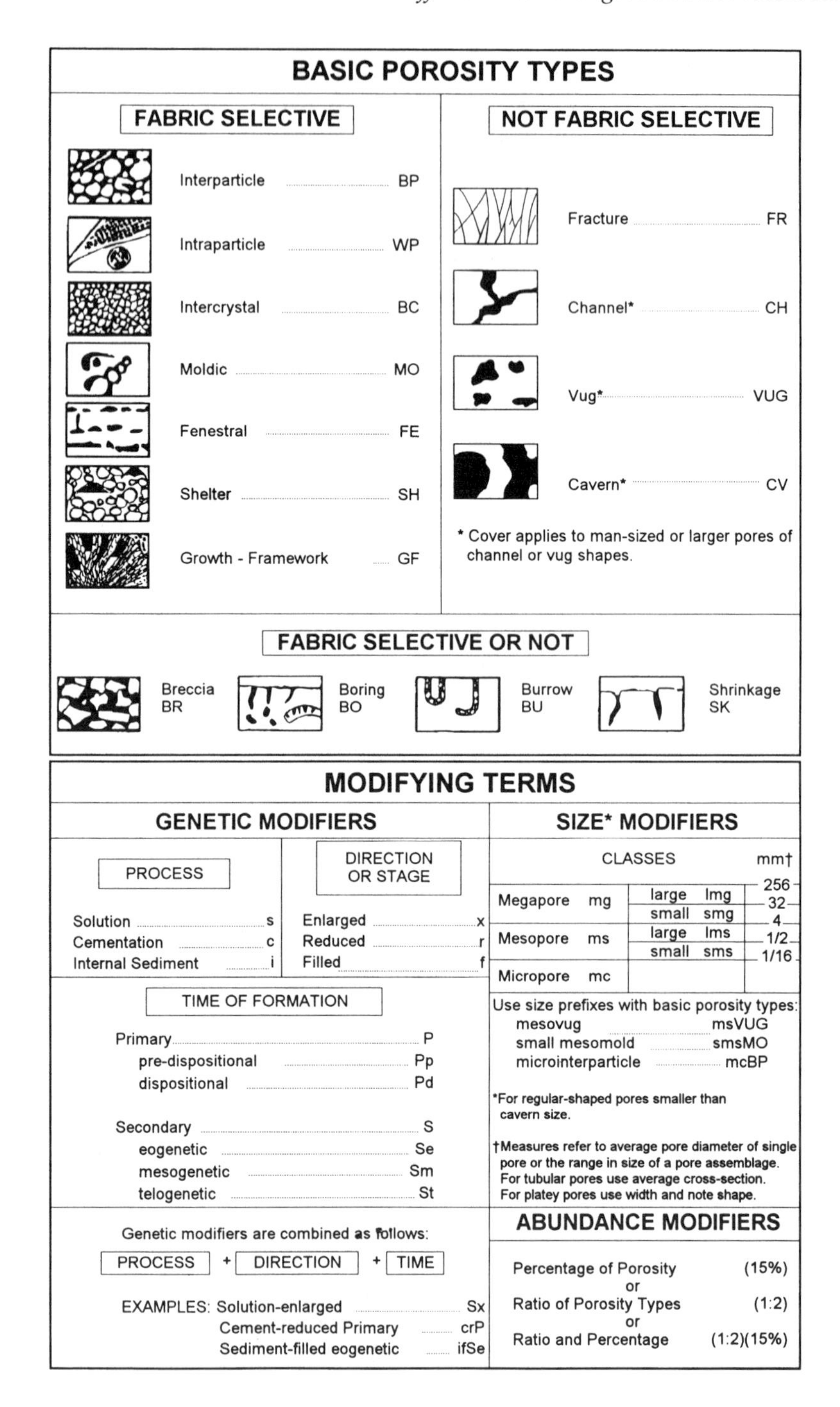

Figure E.3 Classification of porosity types. (After Choquette and Pray, 1970.)

Table E.1 Summary of Diagenesis and Secondary Porosity

Depositional Processes	Diagenetic Processes
Texture Grain Size Sorting Grain Slope Grain Packing Grain Roundness Mineral Composition	Compaction Recrystallization Dissolution Replacement Fracturing Authigenesis Cementation

(Source: After Testa, 1994.)

Specific retention (Sr) is the ratio of the volume of water a soil or rock can retain against gravity drainage compared to the total volume (Vt) of the soil or rock.

$$n = Sr + Sy$$

Total porosity (n) is equal to the specific retention (Sr) plus the specific yield (Sy). Porosity, specific yield, and hydraulic conductivity for various materials are summarized in Table E.1. Average velocity can be calculated for MTBE.

$$Vc = V/R$$

Vc is the average velocity of the contaminant, such as MTBE, V is the average groundwater velocity, and R is the retardation factor. MTBE has a relatively low retardation factor, whereas other chemicals such as the BTEX compounds have retardation factors several times higher. The retardation factor relates to the specific contaminant within an aquifer. For example, if MTBE has a retardation factor of 1.4 and groundwater flow is 5.0 feet per day, then the rate of MTBE transport is about 3.6 feet per day. If, in the same scenario, the retardation for benzene is 8.5, then the transport for benzene is about 0.6 feet per day.

Hydraulic conductivity for various materials is included in Table E.2. The partitioning of volatile organic compounds between the aqueous, vapor, and sorbed phases is shown in Figure E.4. Since MTBE does not move readily into the vapor phase, MTBE generally is found in the saturated zone.

Porosity can be primary or secondary. Primary porosity develops as the sediment is deposited and includes inter- and intraparticle porosity. Secondary porosity develops after deposition or rock formation and is referred to as diagenesis (Choquette and Pray, 1970). Figure E.5 demonstrates petrographic criteria for secondary porosity.

Permeability is a measure of the connectedness of the pores. Thus, a basalt containing many unconnected air bubbles may have high porosity but

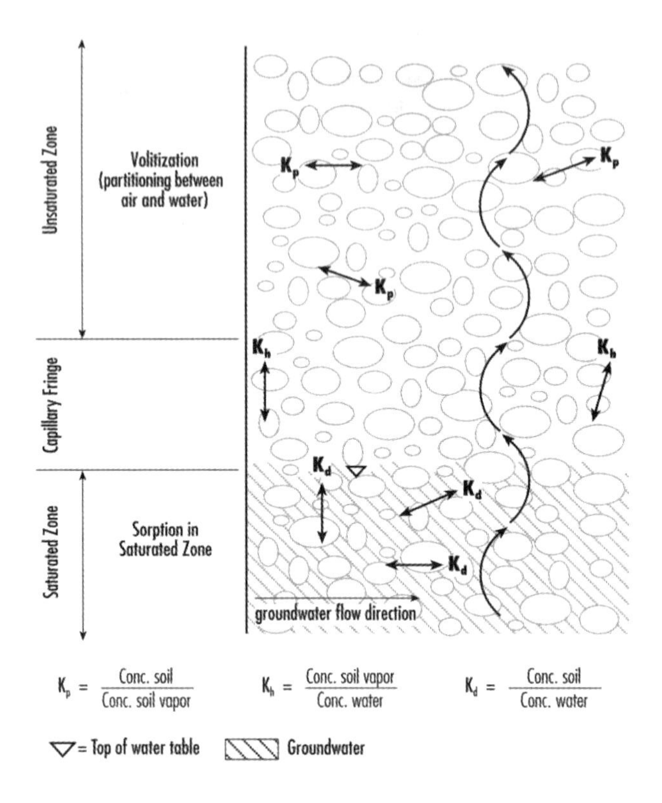

Figure E.4 Equilibrium forces for partitioning of volatile organic compounds between aqueous, vapor, and sorbed phases.

no permeability, whereas a sandstone with many connected pores will have both high porosity and high permeability. Likewise, a fractured, dense basaltic rock may have low porosity but high permeability because of the fracture flow. The nature of the porosity and permeability in any material can change dramatically through time. Porosity and permeability can increase, for example, with the dissolution of cements or matrix, faulting, or fracturing. Likewise, porosity and permeability can decrease with primary or secondary cementation and compaction.

Once a sediment is deposited, diagenetic processes begin immediately and can significantly affect the overall porosity and permeability of the unconsolidated materials. These processes include compaction, recrystallization, dissolution, replacement, fracturing, authigenesis, and cementation (Schmidt, McDonald, and Platt, 1977) (see Table E.1). Compaction occurs by the accumulating mass of overlying sediments called overburden. Unstable minerals may recrystallize, changing the crystal fabric but not the mineral-

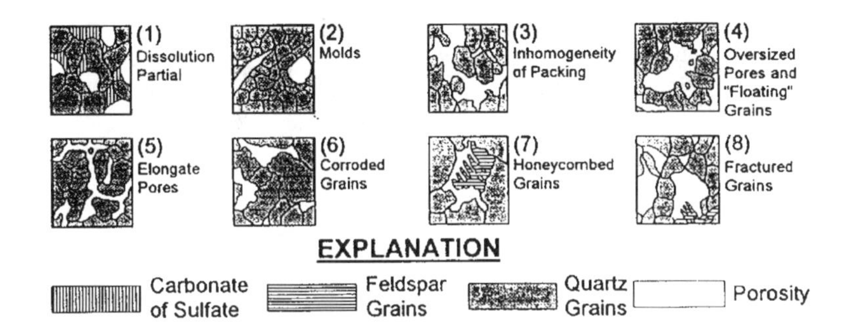

Figure E.5 Petrographic criteria for secondary porosity. (After Schmidt, McDonald, and Platt, 1977.)

Table E.2 Summary of Hydraulic Properties for Certain Lithological Precesses

Process	Effects
Leaching	Increase n and K
Dolomitization	Increase K; can also decrease n and K
Fracturing Joints, Breccia	Increase K; can also increase channeling
Recrystallization	May increase pore size and K; can also decrease n and K
Cementation by calcite, dolomite, anhydrite, pyrobitumen, silica	Decrease n and K

ogy, or they may undergo dissolution and/or replacement by other minerals. Dissolution and replacement processes are common with limestones, sandstones, and evaporites. Authigenesis refers to the precipitation of new mineral within the pore spaces of a sediment. Lithification occurs when the cementation is of sufficient quantity and the sediment is changed into a rock. Examples of lithification include sands and clays changing into sandstones and shales, respectively.

The most important parameters influencing porosity in sandstone are age (time of burial), mineralogy (i.e., detrital quartz content), sorting, and the maximum depth of burial, and to a lesser degree, temperature. Compaction and cementation will reduce porosity, although porosity reduction by cement is usually only a small fraction of the total reduction (see Figure E.6). The role of temperature probably increases mal gradient of 4°C per 100 m. Uplift and erosional unloading may also be important in the development of fracture porosity and permeability. Each sedimentary and structural basin has its own unique burial history, and the sediments and rocks will reflect unique temperature and pressure curves vs. depth.

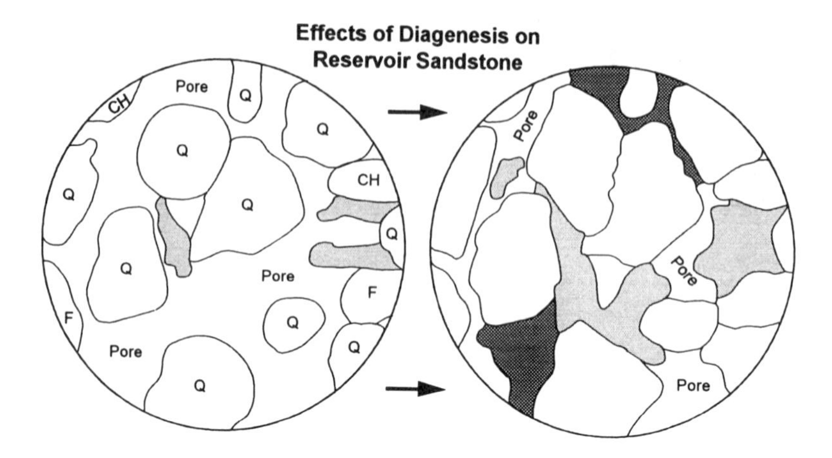

Figure E.6 Reduction in porosity in sandstone as a result of cementation and growth of authigenic minerals in the pores affecting the amount, size, and arrangement of pores. (Modified after Ebanks, 1987.)

E.4 Sedimentary sequences and facies architecture

Analysis of sedimentary depositional environments is important since groundwater resource usage occurs primarily in unconsolidated deposits formed in these environments. Aquifers are water-bearing zones often considered of beneficial use and warrant protection. Most subsurface environmental investigations conducted are also performed in these types of environments.

Erroneous hydraulic or MTBE distribution information can create misinterpretations for several reasons. In actuality, however, (1) the wells were screened across several high-permeability zones or across different zones creating the potential for cross contamination of a clean zone by migrating MTBE from impacted zones, (2) inadequate understanding of soil-gas surveys and minimal MTBE vapor-phase transport, (3) wells screened in upward-fining sequences with the accurately assessing MTBE concentrations within expectation of soil and groundwater within the upper fine-grained section of the sequence, or (4) the depositional environment was erroneously interpreted. Heterogeneities within sedimentary sequences can range from large-scale features associated with different depositional environments that further yield significant large- and small-scale heterogeneities via development of preferential grain orientation. This results in preferred areas of higher permeability and, thus, preferred migration pathways of certain constituents considered hazardous.

To adequately characterize these heterogeneities, it becomes essential that subsurface hydrogeologic assessment include determination of the following:

- Depositional environment and facies of all major stratigraphic units present.

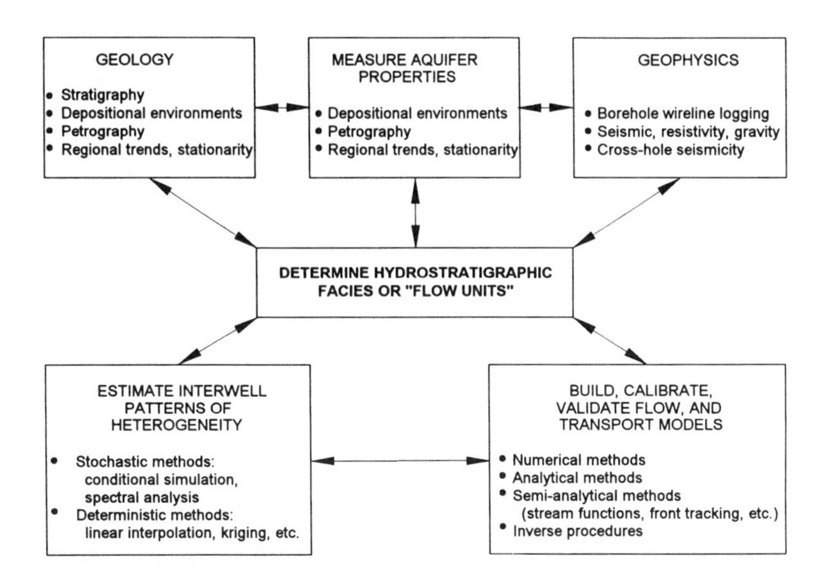

Figure E.7 Schematic depicting the various components of an integrated aquifer description.

- Propensity for heterogeneity within the entire vertical and lateral sequence and within different facies of all major stratigraphic units present.
- Potential for preferential permeability (i.e., within sand bodies).

The specific objectives to understanding depositional environments as part of subsurface environmental studies are to (1) identify depositional processes and resultant stratification types that cause heterogeneous permeability patterns, (2) measure the resultant permeabilities of these stratification types, and (3) recognize general permeability patterns that allow simple flow models to be generated. Flow characteristics in turn are a function of the types, distributions, and orientations of the internal stratification. Since depositional processes control the zones of higher permeability within unconsolidated deposits, a predictive three-dimensional depositional model to assess potential connections or intercommunication between major zones of high permeability should also be developed. A schematic depicting the various components of an integrated aquifer description has been developed, as shown in Figure E.7.

Understanding the facies architecture is extremely important to successful characterization and remediation of contaminated soil and groundwater. Defining a hydrogeologic facies can be complex. Within a particular sedimentary sequence, a hydrogeologic facies can range over several orders of magnitude. Other parameters such as storativity and porosity vary over a range of only one order of magnitude. A hydrogeologic facies is defined as a homogeneous, but anisotropic, unit that is hydrogeologically meaningful for purposes of conducting field studies or developing conceptual models. Facies can be gradational in relation to other facies, with a horizontal length that is finite but usually greater than its corresponding vertical length. A hydrogeologic facies can also be viewed as a sum of all the primary characteristics of a sedimentary unit. A facies can thus be referred to according to one or several factors, such as lithofacies, biofacies, geochemical facies, etc. For example, three-dimensional sedimentary bodies of similar textural character are termed lithofacies. It is inferred that areas of more rapid plume migration and greater longitudinal dispersion correlate broadly with distribution and trends of coarsE.grained lithofacies and are controlled by the coexistence of lithologic and hydraulic continuity. Therefore, lithofacies distribution can be used for preliminary predictions of MTBE migration pathways and selection of a subsurface assessment and remediation strategy. However, caution should be taken in proximal and distal assemblages where certain layered sequences may be absent due to erosion and the recognition of cyclicity is solely dependent on identifying facies based simply on texture. Regardless, the facies reflects deposition in a given environment and possesses certain characteristics of that environment. Sedimentary structures also play a very important role in deriving permeability-distribution models and developing fluid-flow models.

Nearly all depositional environments are heterogeneous, which, for all practical purposes, restricts the sole use of homogeneous-based models in developing useful hydraulic conductivity distributions data for assessing preferred MTBE migration pathways, and developing containment and remediation strategies. There has been much discussion in the literature regarding the influences of large-scale features such as faults, fractures, significantly contrasting lithologies, diagenesis, and sedimentalogical complexities. Little attention, however, has been given to internal heterogeneity within genetically defined sand bodies caused by sedimentary structures and associated depositional environment and intercalations. In fact, for sand bodies greater variability exists within bedding and lamination pockets than between them. An idealized model of the vertical sequence of sediment types by a meandering stream shows the highest horizontal permeability (to groundwater or MTBE) to be the cross-bedded structure in a fine to medium grained, well-sorted unconsolidated sand or sandstone (see Figure E.8).

The various layers illustrated affect the flow of fluids according to their relative characteristics. For example, in a point-bar sequence, the combination of a ripple-bedded, coarser-grained sandstone will result in retardation of flow higher in the bed, and deflection of flow in the direction of dip of the lower trough crossbeds. See Table E.2 for a summary of hydraulic properties for certain depositional environments.

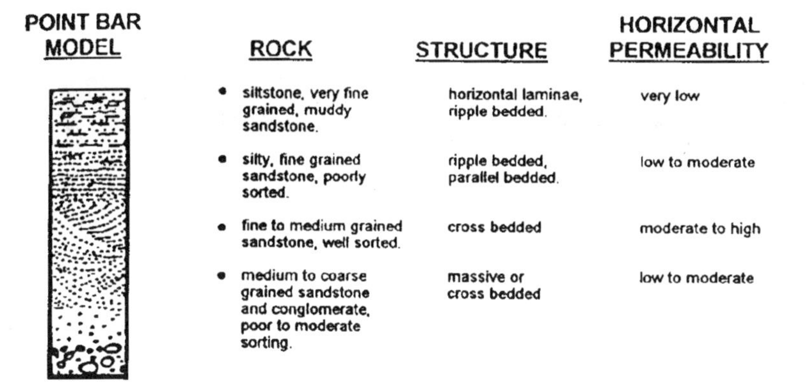

Figure E.8 Point bar geologic model showing the influences of a sequence of rock textures and structures in an aquifer consisting of a single point bar deposit on horizontal permeability excluding effects of diagenesis. (Modified after Ebanks, 1987.)

Hydrogeologic analysis is conducted in part by the use of conceptual models. These models are used to characterize spatial trends in hydraulic conductivity and permit prediction of the geometry of hydrogeologic facies from limited field data. Conceptual models can be either site specific or generic. Site-specific models are descriptions of site-specific facies that contribute to understanding the genesis of a particular suite of sediments or sedimentary rocks. The generic model, however, provides the ideal case of a particular depositional environment or system. Generic models can be used in assessing and predicting the spatial trends of hydraulic conductivity and, thus, dissolved MTBE in groundwater. Conventional generic models include either a vertical profile that illustrates a typical vertical succession of facies, or a block diagram of the interpreted three-dimensional facies relationships in a given depositional system.

Several of the more common depositional environments routinely encountered in subsurface environmental studies are discussed below. Included is discussion of fluvial, alluvial fan, glacial, deltaic, eolian, carbonate, and volcanic-sedimentary sequences. Hydrogeologic parameters per depositional environment are available from the literature.

E.4.1 Fluvial sequences

Fluvial sequences are difficult to interpret due to their sinuous nature and the complexities of their varied sediment architecture reflecting complex depositional environments (see Figure E.9). Fluvial sequences can be divided into high-sinuosity meandering channels and low-sinuosity braided channel complexes. Meandering stream environments (i.e., Mississippi River) consist of an asymmetric main channel, abandoned channels, point bars, levees, and floodplains.

Table E.3 Summary of Deltaic Sequences and Characteristics

Depositional Environment	Hydrogelogic Facies	Hydraulic Conductivity[b,c]		Porosity in percent[c]	Ref.
		Horizontal	Vertical		
Eolian	Dune sand	5-140 (54)		42-55 (49)	Pryor (1973)
	Interdune/extra-erg	0.67-1,800			Chandler et al. (1989)
	Wind-ripple	900-5,200			Chandler et al. (1989)
	Grain flow	3,700-12,000			Chandler et al. (1989)
Fluvial	River point bar	4-500 (93)		17-52 (41)	
	Beach sand	3.6-166 (68)		39-56 (49)	
Glacial	Meltwater streams	10^{-1}-10^{-5}cm/s			Anderson (1989)
	Outwash drift	10^{-3}-10^{-4} [d]	10-11[e]		Sharp (1984)
	Basal till	10^{-4}-10^{-9}cm/s			Anderson (1989)
	Ester sediment	10^{-1}-10^{-3}cm/s			Caswell (1988a; 1988b); DeGear (1986); Patson (1970)
	Supraglacial sediments	10^{-3}-10^{-7}cm/s			Stephanson et al. (1989)
Deltaic	Distributary channel sandstone	(436)		(28)	Tillman and Jordan (1987)
	Splay channel sandstone	(567)		(27)	Tillman and Jordan (1987)
	Wave-dominated sandstone within prodelta and shelf mudstone	(21)		(21)	Tillman and Jordan (1987)
Volcanic-Sedimentary	Basalt (CRG)[f]	0.002-1,600 (0.65)	10^{-8}-10		Lindholm and Vaccaro (1988)
	Basalt (SRG)[f]	150-3,000			Lindholm and Vaccaro (1988)
	Basalt (CRG)	1x10^{-8}-10^{-5}cm/s	1x10^{-7}-2x10^{-7}cm/s		Testa (1988); Wang and Testa (1989)
	Sedimentary Interbed (SRG)	3x10^{-6}-3x10^{-2} [e]			Lindholm and Vaccaro (1988)
	Tuffaceous siltstone (interbeds; CRG)	1x10^{-6}-2x10^{-4}cm/s	1x10^{-8}-1x10^{-3}cm/s[e]	27-68 (42)	Testa (1988); Wang and Testa (1989)
	Interflow zone (CRG)	2x10^{-4}cm/s			Testa (1988); Wang and Testa (1989)

[a] Carbonates not represented but can have permeabilities ranging over 5 orders of magnitude.
[b] Values are in millidarcy (mD) per day unless otherwise noted; cm/s = centimeters per second; 1 mD = .001 darcy; 1 cm/s = 1.16x10^{-3} darcy.
[c] Values shown in parentheses are averages.
[d] Field.
[e] Laboratory
[f] CRG = Columbia River Group; CRG = Snake River Group

Usually developed where gradients and discharge are relatively low, five major lithofacies have been recognized:

- Muddy fine-grained streams.
- Sand-bed streams with accessory mud.
- Sand-bed streams without mud.
- Gravelly sand-bed streams.
- Gravelly streams without sand.

Meandering streams can also be subdivided into three subenvironments: floodplain subfacies, channel subfacies, and abandoned channel subfacies. Floodplain subfacies is comprised of very fine sand, silt, and clay deposited

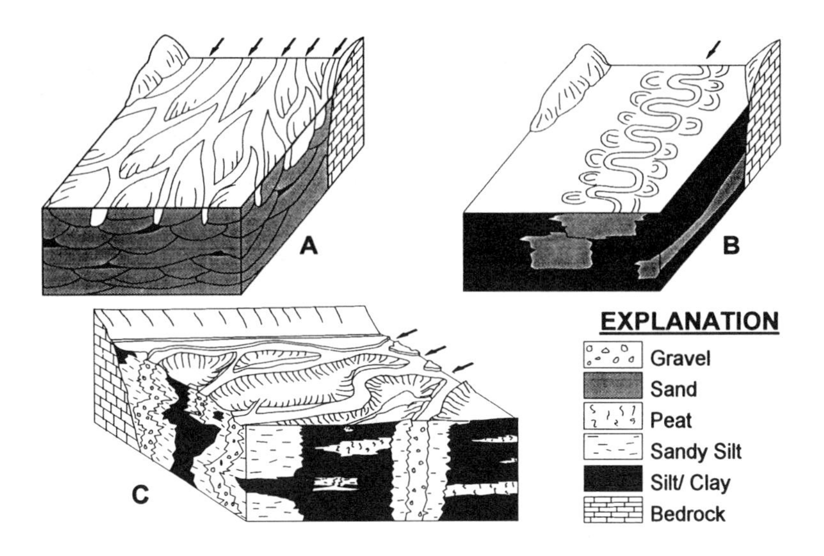

Figure E.9 Fluvial facies model illustrating contrasting patterns of heterogenity in (a) braided rivers, (b) meandering rivers, and (c) anatomosing rivers. (Modified after Allen, 1965 and Smith and Smith, 1980.)

on the overbank portion of the floodplain, out of suspension during flooding events. Usually laminated, these deposits are characterized by sand-filled shrinkage cracks (subaerial exposure), carbonate caliches, laterites, and root holes. The channel subfacies is formed as a result of the lateral migration of the meandering channel which erodes the outer concave bank, scours the riverbed, and deposits sediment on the inner bank referred to as the point bar. Very characteristic sequences of grain size and sedimentary structures are developed. The basal portion of this subfacies is lithologically characterized by an erosional surface overlain by extraformation pebbles and intraformational mud pellets. Sand sequences with upward fining and massive, horizontally stratified and trough cross-bedded sands overlie these basal deposits. Overlying the sand sequences are tabular, planar, cross-bedded sands which grade into microcross-laminated and flat-bedded fine sands, grading into silts of the floodplain subfacies. The abandoned channel subfacies are curved fine-grained deposits of infilled abandoned channels referred to as oxbow lakes. Oxbow lakes form when the river meanders back, short-circuiting the flow. Although lithologically similar to floodplain deposits, geometry and absence of intervening point-bar sequences distinguish it from the abandoned channel subfacies.

Braided river systems consist of an interlaced network of low-sinuosity channels and are characterized by relatively steeper gradients and higher discharges than meandering rivers. Typical of regions where erosion is rapid, discharge is sporadic and high, and little vegetation hinders runoff, braided rivers are often overloaded with sediment. Because of this sediment overload, bars are formed in the central portion of the channel around which two new channels are diverted. This process of repeated bar formation and

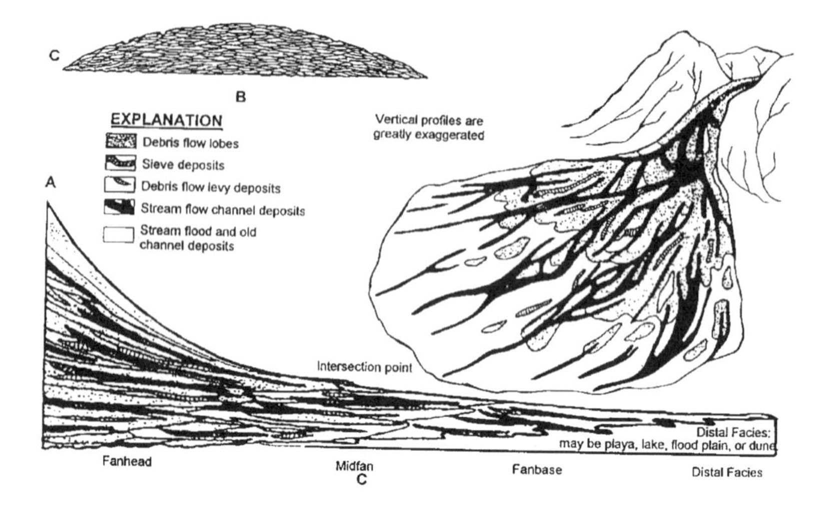

Figure E.10 Generalized model of alluvial fan sedimentation showing (a) fan surface, (b) cross-fan profile, and (c) radial profile. (After Spearing, 1980.)

channel branching generates a network of braided channels throughout the area of deposition.

Lithologically, alluvium derived from braided streams is typically composed of sand and gravel-channel deposits. Repeated channel development and fluctuating discharge results in the absence of laterally extensive cyclic sequences as produced with meandering channels. Fine-grained silts are usually deposited in abandoned channels formed by both channel choking and branching, or trapping of fines from active downstream channels during eddy reversals.

The degree of interconnectedness is important in addressing preferred migration pathways in fluvial sequences. Based on theoretical models of sand–body connectedness, the degree of connectedness increases very rapidly as the proportion of sand bodies increases above 50%. Where alluvial soils contain 50% or more of overbank fines, sand bodies are virtually unconnected.

E.4.2 *Alluvial fan sequences*

Alluvial fan sequences accumulate at the base of an upland area or mountainous area as a result of an emerging stream. These resulting accumulations form segments of a cone with a sloping surface ranging from less than 1° up to 25°, averaging 5°, and rarely exceeding 10°. Alluvial fans can range in size from less than 100 m to more than 150 km in radius, although typically averaging less than 10 km. As the channels shift laterally through time, the deposit develops a characteristic fan shape in plane view, a convex-upward cross-fan profile, and concave-upward radial profile (see Figure E.10).

Facies analysis of alluvial fans requires data on morphology and sediment distribution, and can be divided into four facies: proximal, distal, and, of lesser importance, outer fan and fan fringe facies. Proximal facies are deposited in the upper or inner parts of the fan near the area of stream emergence and are comprised of relatively coarser-grained sediments. The proximal facies comprising the innermost portion of the fan (i.e., apex or fan-head area) contains an entrenched straight valley which extends outward onto the fan from the point of stream emergence. This inner fan region is characterized by two subfacies: a very coarse-grained, broad, deep deposit of one or several major channels, and a finer-grained channel-margin level and interchannel deposit which may include coarse-grained landslides and debris flows type material. Distal facies are deposited in the lower and outer portion of the fan, and are comprised of relatively finer-grained sediments. The distal facies typically comprises the largest area of most fans, and consists of smaller distributary channels radiating outward and downfan from the inner fan valley. Hundreds of less-developed channels may be present on the fan. Depending on fan gradient, sediment time and supply, and climatic effects among other factors, commonly braided but straight, meandering, and anastomosing channel systems may also be present. Outer fan facies are comprised of finer-grained, laterally extensive, sheet-like deposits of nonchannelized or less-channelized deposits. These deposits maintain a very low longitudinal gradient. The fanfringe facies is comprised of very fine-grained sediments that intertongue with deposits of other environments (i.e., eolian, fluvial, lacustrine, etc.). Most deposits comprising alluvial fan sequences consist of fluvial (streamflow) or debris flow types.

Alluvial fans are typically characterized by high permeability and porosity. Groundwater flow is commonly guided by paleochannels which serve as conduits, and relatively less permeable and porous debris and mud flow deposits. The preponderance of debris flow and mudflow deposits in the medial portion of fans may result in decreased and less-predictable porosity and permeability in these areas. Aquifer characteristics vary significantly with the type of deposit and relative location within the fan. Pore space also develops as intergranular voids, interlaminar voids, bubble cavities, and desiccation cracks.

E.4.3 Deltaic sequences

Deltas are abundant throughout the geologic record with 32 large deltas forming at this time and countless others in various stages of growth. A delta deposit is partly subaerial built by a river into or against a body of permanent water. Deltaic sedimentation requires a drainage basin for a source of sediment, a river for transport of the material, and a receiving basin to store and rework it. During formation, the outer and lower parts are constructed below water, and the upper and inner surfaces become land reclaimed from the sea. Deltas form by progradation or a building outward of sediments onto themselves (see Figure E.11).

As the delta system progrades further and further, the slope and discharge velocity lessens and the carrying capacity of the delta is reduced by

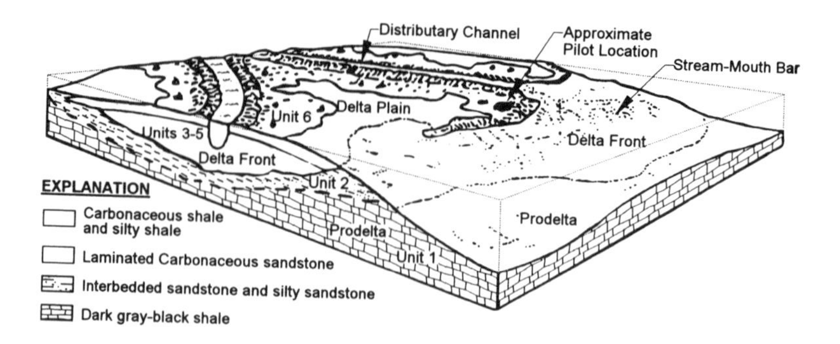

Figure E.11 Block diagram showing vertical and aerial distribution of units in a typical modern delta. (Modified after Harris, 1975.)

its own sediment load. Once a particular pathway in a delta system is no longer available due to the sediment buildup and upward vertical migration of the channel bed and the adjacent levees, another delta system forms in a different location, usually by a break in a levee wall in an upgradient position. This break in the wall and the shift in the locus of deposition increase the slope, sediment-carrying capacity, and discharge velocity of the new delta system. Any given delta is thus a composite of conditions reflecting initiation of delta development to its ultimate abandonment of a particular deposition center.

Delta sequences reflect condition of source (volume and type of available sediment) and distribution and dispersal processes. Two general classes or end members have been defined: high- and low-energy deltas. High-energy deltas or sand deltas are characterized by few active meandering distributary channels, with the shoreline comprised of continuous sand (i.e., Nile, Rhone, or Brazos-Colorado). Low-energy or mud deltas are characterized by numerous bifurcating or branching, straight to sinuous, distributary channels, with shorelines comprised of discontinuous sands and muds.

No two deltas are exactly alike in their distribution and continuity of permeable and impermeable sediments. The most important parameters controlling the distribution are size and sorting of grains. All delta systems form two parts during development: a regressive sequence of sediment produced as the shoreline advances seaward and a system of distributary channels. These two parts result in two main zones of relatively high permeability: channel sands and bar sands.

Typical deltaic sequences from top to bottom include marsh, inner bar, outer bar, prodelta, and marine (see Table E.3). Depositional features include distributary channels, river-mouth bars, interdistributary bays, tidal flats and ridges, beaches, eolian dunes, swamps, marshes, and evaporite flats (see Figure E.12).

The geometry of channel deposits reflects delta size, position of the delta in the channel, type of material being cut into, and forces at the mouth of the channel distributing the sediments. For high-energy sand deltas, channels can be filled with up to 90% sand, or clay and silt; for example, those sequences of high permeability in bar deposits at or near the top with

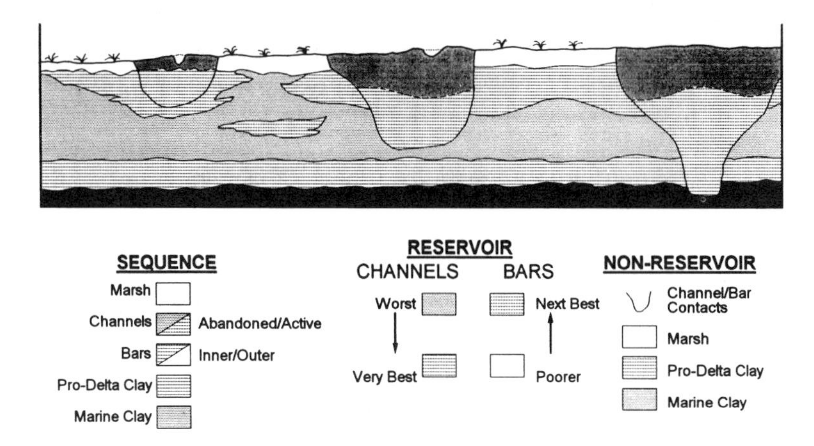

Figure E.12 Deltaic facies model illustrating heterogenity and continuity. (After Testa, 1994.)

decreasing permeability with depth and laterally away from the main sand buildup. Within bar deposits, permeability generally increases upward and is highest at or near the top, and is similar parallel to the trend of the bar (i.e., highest at the top near the shore, decreasing progressively seaward). Porosity is anticipated to be well connected throughout the bar with the exception of the lower part. In channels, however, high-permeability sequences occur in the lower part with decreasing permeability vertically upward. Within channel deposits, porosity and permeability are high at the lower part of the channel, decreasing vertically upward, with an increase in the number, thickness, continuity, and aerial extent of clay interbeds.

Low-energy mud deltas differ from sand deltas in that sediments are carried into a basin via numerous channels during flooding events, and the fine-grained silts and clay are not winnowed from the sand before new sediment circulates from the next flooding event. The coarse-grained sediment is thus more discontinuous with more numerous (less than 1 inch thick) and continuous clay and silt interbeds. Along the perimeter or within bar deposits, grain size increases with sorting and improves vertically upward; clay and silt interbeds decrease in number, thickness, and aerial extent. Within individual bars, overall permeability decreases laterally away from the coarse-grained sand depositional pathways. The highest permeability is at or near the top of the bar decreasing vertically downward and in a seaward direction. Sand continuity, thus permeability, is poor due to the numerous shifting distributary channels forming widespread clay and silt interbeds ranging from less than 1 foot to more than 12 feet in thickness. Coarse-grained sands predominate within the lower portion of the distributary channels with clay and silt interbeds typically less than 1 foot in thickness, and range from a few feet to a few tens of feet in maximum aerial extent, thus not providing a barrier to vertical flow. The number, thickness, and aerial extent of these fine-grained interbeds generally increase vertically upward depending on how fast and where the channel was abandoned. Overall, permeability and porosity continuity are high only in the upward portion of the bar. Within the channels, however, permeability and porosity continuity are high at the basal portion of the channels, but the amount and quality of coarse-grained sand (high permeability zones) are dependent upon the location and rate of channel abandonment.

Average porosity and permeability based on a broadly lenticular wave-dominated deltaic sandstone (e.g., Upper Cretaceous Big Wells aquifer, which is one of the largest oil fields located in south Texas) increased in prodelta and shelf mudstones, averaging 21% and 6 mD, respectively. Studies of the El Dorado field located in southeastern Kansas, a deltaic sequence containing the 650-feet-thick Admire sandstone, have reported porosity and permeability averaging 28% and 436 mD, respectively, within the distributary channel sandstones. Thinner and discontinuous splay channels sandstones average 27% porosity and 567 mD in permeability. The variation in porosity and permeability reflects diagenetic processes (i.e., deformation, secondary leaching of feldspar, and formation of calcite cement and clay laminae).

E.4.5 Glacial sequences

Sequences derived from glacial processes include four major types of materials: tills, ice-contact, glacial fluvial or outwash, and delta and glaciolacustrine deposits. Glacial tills make up a major portion of a group of deposits referred to as diamictons which are defined as poorly sorted, unstratified deposits of nonspecific origin. Tills and associated glaciomarine drift deposits are both deposited more or less directly from ice without the winnowing effects of water. Till is deposited in direct contact with glacial ice and, although substantial thickness accumulations are not common, tills make up a discontinuous cover totaling up to 30% of the earth's continental surface.

Glaciomarine drift, however, accumulates as glacial debris melts out of ice floating in marine waters. These deposits are similar to other till deposits but also include facies that do not resemble till or ice-contact deposits. A lesser degree of compaction is evident due to a lack of appreciable glacial loading.

Tills can be divided into two groups based on deposition (basal till or supra glacial till), or three groups based upon physical properties and varying depositional processes: lodgement, ablation, and flow. Lodgement tills are deposited subglacially from basal, debris-laden ice. High shear stress results in a preferred fabric (i.e., elongated stones oriented parallel to the direction of a flow) and a high degree of compaction, high bulk densities, and low void ratios of uncemented deposits. Ablation tills are deposited from englacial and superglacial debris dumped on the land surface or the ice melts away. These deposits lack significant shear stresses and thus are loosely consolidated with a random fabric. Flow tills are deposited by water-saturated debris flowing off glacial ice as mudflow. Flow tills exhibit a high

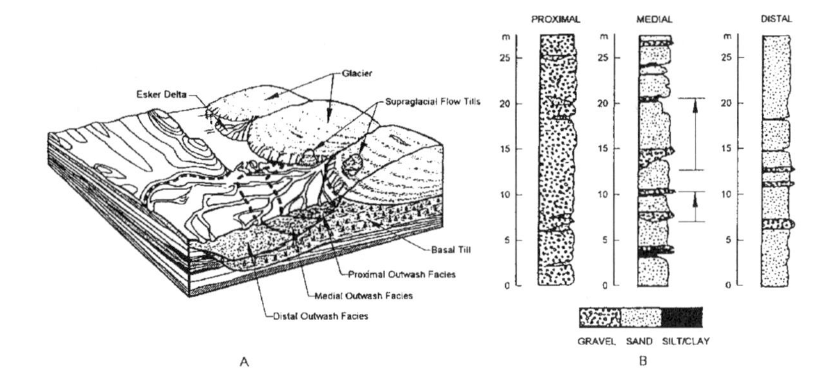

Figure E.13 Hydrogeologic facies model for glacial depositional environment. (After Testa, 1994.)

degree of compaction, although less than that of lodgement tills, with a preferred orientation of elongated stones due to flowage (see Figure E.13).

Till is characterized by a heterogeneous mixture of sediment sizes (boulders to clay) and a lack of stratification. Particle size distribution is often bimodal with predominant fractions in the pebble-cobble range and silt-clay range, both types being massive with only minor stratified intercalations. Other physical characteristics of till include glaciofluvial deposits or outwash deposits having strong similarities to sediments formed in fluvial environments due to similar transportation and deposition mechanisms. These types of deposits are characterized by abrupt particle-size changes and sedimentary structures reflecting fluctuating discharge and proximity to glaciers. Characteristics include a downgradient fining in grain size and down-gradient increase in sorting, therefore a decrease in hydraulic conductivity. Outwash deposits can be divided into three facies: proximal, intermediate or medial, and distal. Outwash deposits are typically deposited by braided rivers, although the distal portions are deposited by meandering and anastomosing rivers. Proximal facies are deposited by gravel-bed rivers while medial and distal facies are deposited by sand-bed rivers. Thus, considerable small-scale variability within each facies assemblage exists. Vertical trends include fining-upward sequences as with meandering fluvial sequences. Within the medial portion, series of upward fining or coarsening cycles are evident depending on whether the ice front was retreating or advancing, respectively. Layered sequences within the gravel-dominant proximal facies and sand-dominant distal facies are either absent or hindered by the relatively large-grain size component of the proximal facies. A hydrogeologic facies model and respective vertical profiles has been developed.

Delta and glaciolacustrine deposits are formed when meltwater streams discharge into lakes or seas. Ice-contact delta sequences produced in close proximity to the glacier margins typically exhibit various slump-deformation structures. Delta sequences produced a considerable distance from the glacier margins exhibit no ice-collapse structures, variable sediment discharge, and particle-size distribution and structures (i.e., graded bedding, flow rolls, varies, etc.) similar to that of meltwater streams.

Also associated with till deposits are ice-contact deposits which form from meltwater on, under, within, or marginal to the glacier. Detritus deposits formed on, against, or beneath the ice exhibit better sorting and stratification, a lack of bimodal particle-size distribution, and deformational features such as collapse features (i.e., tilting, faulting, and folding).

Hydrogeologically, hydraulic conductivity of basal tills facies on the order of 10^{-4} cm/s with horizontal hydraulic conductivities on the order of 10^{-3} to 10^{-7} cm/s reflecting locations and degree of interconnected sand and gravel channel deposits contained within the till. Drift deposits can vary from about 10^{-11} m/s (laboratory tests) to 10^{-6} to 10^{-7} (field) when permeable sand lenses or joints are intersected.

E.4.5 Eolian sequences

Eolian or wind-deposited sediments are complex, highly variable accumulations. They are characterized as well-sorted, matrix-free, well-rounded sediments with a dominance of sand-sized fractions, and are perceived as essentially lithologically homogeneous with irregular plan and cross-sectional geometrics with the exception of the linear trends of coastal dunes. Unlike many other sedimentary facies, eolian deposits have no predictable geometry and/or cyclic motif of subfacies. Recent studies have provided a better understanding of the stratigraphic complexity and thus flow regime within these deposits (see Figure E.14).

Small-scale forms of eolian deposits include wind sand ripples and wind granule ripples. Wind sand ripples are wavy surface forms on sandy surfaces whose wavelength depends on wind strength and remains constant with time. Wind granule ripples are similar to wind sand ripples but are usually produced in areas of erosion. Excessive deflation produces a large concentration of grains 1 to 3 mm in diameter that are too big to be transported via saltation under the existing wind conditions.

Larger-scale eolian sand forms include sand drifts, sand shadows, gozes, sand sheets, and sand dunes. Sand drifts develop by some fixed obstruction which lies in the path of a sand-laden wind. When sand accumulates in the lee of the gap between two obstacles, a tongue-shaped sand drift develops. As the wind velocity is reduced by an obstruction, a sand shadow develops. Gozes are gentle large-scale undulatory sand surfaces associated with sparse desert vegetation. Sand sheets are more or less flat, with slight undulations or small dune tile features, and encompassing large areas.

Dune Type	Definition and Occurence
Barchan Dune	A crescent-shaped dune with horns pointing downward. Occurs on hard, flat floors of desserts. Constant wind and limited sand supply. Height 1m to more than 30m. 
Transverse Dune	A dune forming a asymmetrical ridge transverse to wind direction. Occurs in area with abundant sand and little vegetation. In places grades into barchans. 
Parabolic Dune	A dune of U-shape with the open end of the U facing upwind. Some form by piling of sand along leeward and lateral margins of areas of deflation in older dunes. 
Linear Dune	A long, straight, ridge-shaped dune parallel with wind direction. As much as 100m high and 100km long. Occurs in deserts with scanty sand supply and strong winds varying within one general direction. Slip faces vary as wind shifts direction. 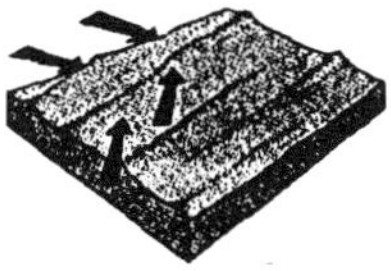
Star Dune	An isolated hill of sand having a base that resembles a star in plan view. Ridges converge from basal points to a central peak as high as 100m. Tends to remain fixed in place in an area where wind blows from all directions. 

Figure E.14 Dune types based on form. (Modified after McKee, 1979.)

Sand dunes are the most impressive features and develop whenever a sand-laden wind deposits in a random patch. This patch slowly grows in height as a mound, until finally a slip face is formed. The sand mound migrates forward as a result of the advance of the slip face, but maintains its overall shape providing wind conditions do not change. Sand dunes are characterized by wind conditions, sand type, and sand supply (see Figure E.14).

Eolian deposits are stratigraphically complex because of (1) differing spatial relationships of large-scale forms such as dunes, interdunes, and sand-sheet deposits relative to one another and to ectradune (noneolian) sediments; and (2) varying dune types, each with its own cross-bedding

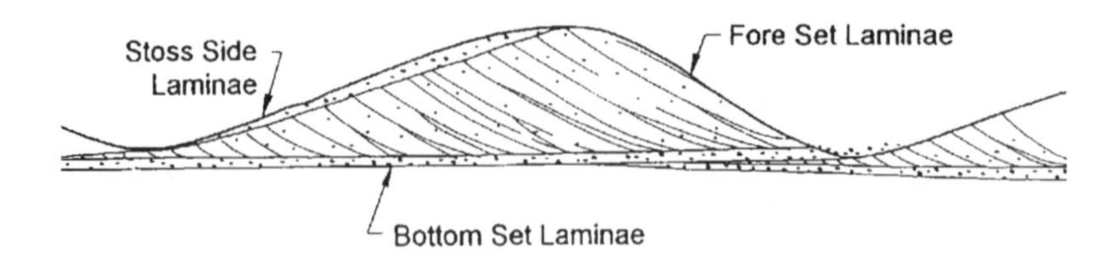

Figure E.15 Basic Eaolian bed forms as related to the number of slip facies. (After Reinick and Singh, 1975.)

patterns and different degrees of mobility; thus, there are different fluid-flow properties when consolidated or lithified. Sedimentary structures within eolian deposits include ripples, contorted sand bedding, cross-bedding with great set heights, normally graded beds, inversely graded beds, evenly laminated beds, discontinuously laminated beds, nongraded beds, and lag deposits along boundary surfaces and sets. Basic eolian bed forms as related to a number of slip facies are well documented (see Figure E.15).

Relatively recent eolian deposits are presumed to have high porosity and permeability, and are typically well rounded, well sorted, and generally only slightly cemented. Regional permeability is usually good due to a lack of fine-grained soils, shales, interbeds, etc., and thus constitutes important aquifers. Studies conducted on several large eolian deposits (i.e., Page sandstone of northern Arizona) have shed some light on preferred fluid migrations in such deposits. For example, fluid flow is directionally dependent because of inverse grading within laminae. Permeability measured parallel to wind-ripple laminae has been shown to be from two to five times greater than that measured perpendicularly to the laminae. Four common cross-set styles are based on bulk permeability and directional controls of each stratum type for the Page sandstone in northern Arizona (see Figure E.16).

Page sandstone is poorly cemented, has high porosity, and has a permeability outcrop which has never been buried. In Figure E.16 Case A, the cross set is composed exclusively of grain-flow strata; thus, the permeability of each grain-flow set is high in all directions with significant permeability contrasts, flow barriers, or severely directional flow. Due to inverse grading, fluid flow is greater parallel to the grain-flow strata than across it. In Case B, laminae occurring in wind-ripple cross sets have low permeability throughout, thus inhibiting flow and imparting preferred flow paths. Case C illustrates bulk permeability based on the ratio of grain-flow strata to wind-ripple strata in the cross set. Higher ratios are indicative of a greater capacity to transmit fluids, while low permeability logs created by wind-ripple sets act to orient fluid migration parallel to the stratification. Case D shows that a cross set which exhibits grain-flow deposits grading into wind-ripple laminae is more permeable at the top of the set due to the dominance of grain-flow strata. Fluid flow is thus reduced downward throughout the set due to the transition from high-permeability grain-flow cross strata to low perme-

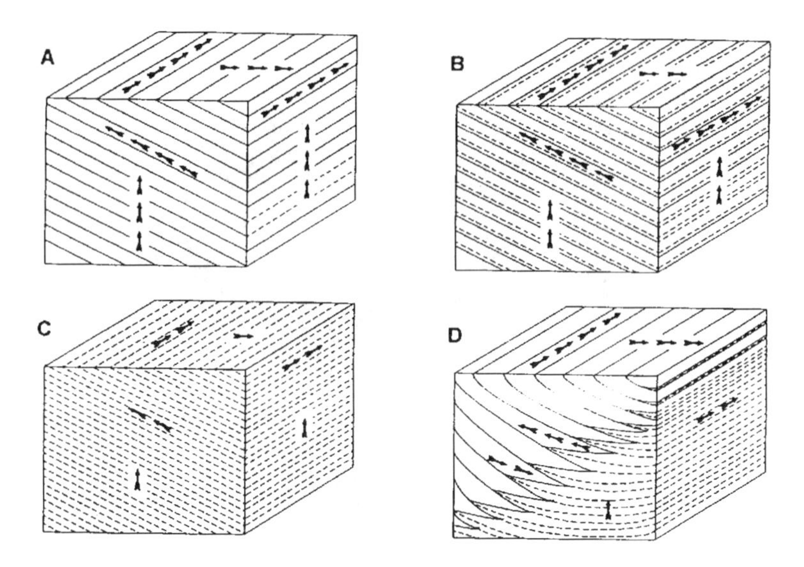

Figure E.16 Common styles of cross-strata in Page sandstone: (a) grain-flow set, (b) wind-ripple set, (c) interlaminated grain-flow and wind-ripple set, and (d) grain-flow foresets toeing into wind-ripple bottom sets. Directional permeability indicated by arrows; more arrows denote higher potential flow. (After Reinick and Singh, 1975.)

ability wind-ripple laminae, with greater ease of flow occurring in the cross set from the wind-ripple laminae into the grain-flow strata (see Figure E.17).

Overall, interdune or extra-erg deposits are least permeable (0.67 to 1800 mD), wind-ripple strata moderately permeable (900 to 5200 mD), and grainflow strata the most permeable (3700 to 12,000 mD).

Compartmentalization develops due to bounding surfaces between the cross sets, with flow largely channeled along the sets. Fluid flow is especially great where low permeability interdune or extra-erg deposits overlie bounding surfaces (i.e., along sets). Flow windows, however, occur where low-permeability strata were eroded or pinched out. Because of high-permeability grain-flow deposits relative to wind-ripple strata, fluid migration between adjacent grain-flow sets would be more rapid than across bounding surfaces separating sets of wind-ripple deposits. Flow through the sets themselves would have been dictated by internal stratification types.

E.4.6 Carbonate sequences

Carbonate sequences are important in that aquifers within such sequences are often heavily depended upon for drinking water, irrigation, and other uses. Carbonate rocks are exposed on over 10% of the earth's land area. About 25% of the world's population depends on fresh water retrieved from Karst aquifers. The Floridan aquifer of Cenozoic age, for example, is the

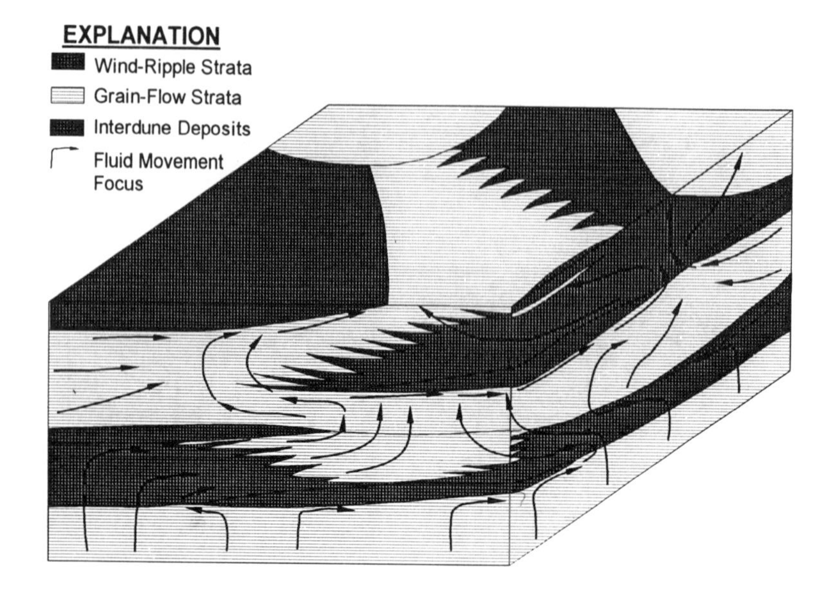

Figure E.17 Fluid flow through idealized eolian sequences based on relative permeability values of stratification types and bounding surfaces, assuming a vertical pressure field. (After Chandler et al., 1989.)

principal aquifer in the southeastern U.S., and is encountered in Florida, Georgia, Alabama, and South Carolina. However, although its major resource is as a potable water supply, the nonpotable part of the aquifer in southern Florida serves as a disposal zone for municipal and industrial wastewater via injection wells.

Karst terrains are the foremost examples of groundwater erosion. Karst terrains can be divided into two types: well developed and incipient. Well-developed karst terrains are marked by surface features such as dolines or sinkholes which can range from 1 to 1000 m in maximum dimension. Other features are closed depressions, dry valleys, gorges, and sinking streams and caves, with local groundwater recharge via both infiltration and point sinks. The subsurface systems of connected conduits and fissure openings, enlarged by solution (i.e., cave systems), serve a role similar to that of stream channels in fluvial systems, and reflect the initial phases of subsequent surface karst landscape features (i.e., sinkholes and depressions). Cave systems actually integrate drainage from many points for discharge at single, clustered, or aligned springs. Incipient karst terrains differ from well-developed terrains in that few obvious surficial features exist and recharge is limited primarily to infiltration. Several other approaches to classification of karst terrains exist: holokarst, merokarst or fluviokarst, and parakarst. In holokarst terrains, all waters are drained underground, including allogenic streams (i.e., those derived from adjacent nonkarst rocks), with little or no

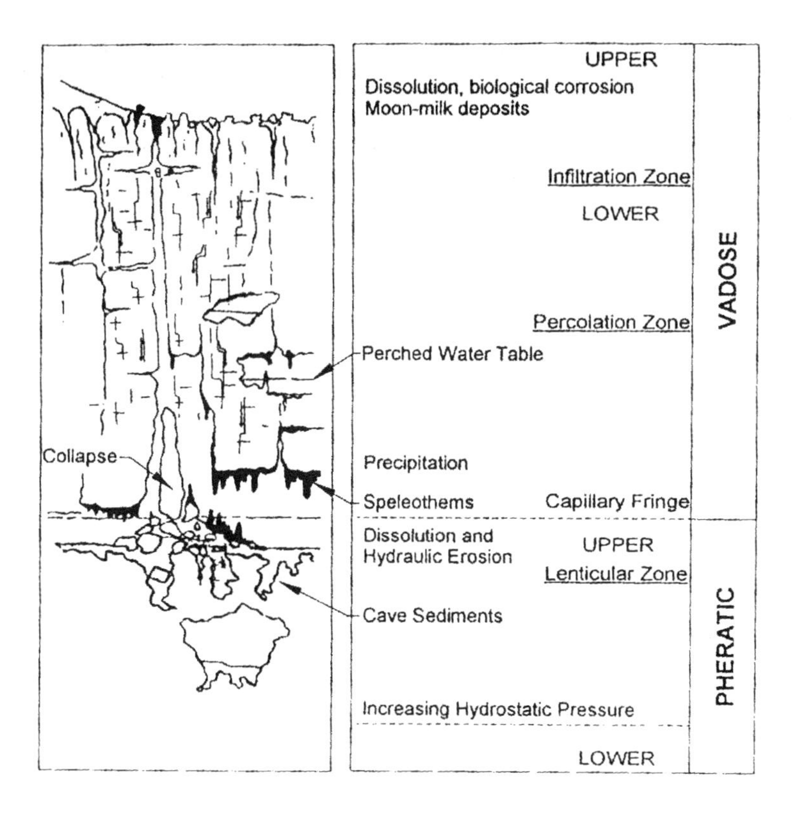

Figure E.18 Idealized authigenic karst profile. (After Scholle et al., 1983.)

surface channel flow. In fluviokarst terrains, major rivers remain at the surface, reflecting either large flow volumes that exceed the aquifer's ability to adsorb the water or immature subsurface development of underground channels. Parakarst terrains are a mixture of the two, reflecting mixtures of karst and nonkarst rocks. Covered karst reflects the active removal of carbonate rocks beneath a cover of other unconsolidated rocks (i.e., sandstone, shales, etc.); whereas, mantled karst refers to deep covers of unconsolidated rocks or materials. Paleokarst terrains are karst terrains or cave systems that are buried beneath later strata and can be exhumed or rejuvenated. Pseudokarst refers to karst-like landforms that are created by processes other than rock dissolution (i.e., thermokarst, vulcanokarst, and mechanical piping). See Figure E.18 for an idealized authigenic karst profile.

Carbonate sediments can accumulate in both marine and nonmarine environments. The bulk of carbonate sediments are deposited in marine environments, in tropical and subtropical seas, with minimal or no influx of terrigenous or land-derived detritus. Marine depositional environments include tidal flat, beach and coastal dune, continental shelf, bank, reef, basin margin and slope, and deeper ocean or basin. Lakes provide the most extensive carbonate deposits on land, regardless of climate, although carbonate can occur or caliche (i.e., soil-zone deposits) and travertine (i.e., caves, karst, and hot-spring deposits).

Carbonate rocks are defined as containing more than 50% carbonate minerals. The most common and predominant carbonate minerals are calcite ($CaCO_3$) and dolomite [$CaMg(CO_3)_2$]. Other carbonate minerals include aragonite ($CaCO_3$), siderite ($FeCO_3$), and magnesite ($MgCO_3$). The term limestone is used for those rocks in which the carbonate fraction is composed primarily of calcite, whereas the term dolomite is used for those rocks composed primarily of dolomite.

Overall, carbonate rocks serve as significant aquifers worldwide and are not limited by location or age of the formation. Carbonate rocks show a total range of hydraulic conductivities over a range of 10 orders of magnitude. The broad diversity in hydrogeologic aspects of carbonate rocks reflects the variable combination of more than 60 processes and controls. Hydrogeologic response is related to rock permeability, which is affected most by interrelated processes associated with dynamic freshwater circulation and solution of the rock. Dynamic freshwater circulation is controlled and maintained primarily by the hydraulic circuit: maintenance of the recharge, flowthrough, and discharge regime. Without these regimes, the overall system is essentially stagnant and does not act as a conduit. The primary controls on solutions include rock solubility and chemical character of the groundwater; secondary controls include diagenetic, geochemical, and chronologic aspects.

Carbonate aquifers are characterized by extremely heterogeneous porosity and permeability, reflecting the wide spectrum of depositional environments for carbonate rocks and subsequent diagenetic alteration of the original rock fabric. Pore systems can range from thick, vuggy aquifers in the coarse-grained skeletal-rich facies of reef or platform margin, to highly stratified, often discontinuous aquifers in reef and platform interiors, and nearshore facies. Due to the brittle nature of limestones and dolomites, most exhibit extensive joint or fault systems because of uneven isostatic adjustment and local stresses produced by solution effects and erosion.

Rainwater commonly absorbs carbon dioxide from the air and forms carbonic acid, a weak acid. Once exposed at or near the surface, limestone and dolomite can be easily dissolved by acidic rainwater (Driscoll, 1986).

Karst terrains are highly susceptible to groundwater contamination. When used for waste disposal, these areas are susceptible to potential failure due to subsidence and collapse, which in turn can result in aquifer compartmentalization. To assess secondary porosity and potential contamination susceptibility, characterization of carbonate or karst aquifers includes generation of data regarding percent rock core recovery, mechanical response during drilling, drilling fluid loss, and drilling resistance.

E.4.7 Volcanic-sedimentary sequences

Volcanic-sedimentary sequences are prevalent in the northwestern U.S. The Columbia Lava Plateau, for example, encompasses an area of 366,000 km^2 and extends into northern California, eastern Oregon and Washington, southern Idaho, and northern Nevada; the Snake River Group encompasses 40,400 km^2 in southern Idaho. The geology of this region consists of a thick, accordantly layered sequence of basalt flows and sedimentary interbeds.

The basalt flows can range from a few 10s of centimeters to more than 100 m in thickness, averaging 30 to 40 m. From bottom to top, individual flows generally consist of a flow base, colonnade, and entablature (see Figure E.19).

The flow base makes up about 5% of the total flow thickness and is typically characterized by a vesicular base and pillow-palagonite complex of varying thickness if the flow entered water. The colonnade makes up about 30% of the flow thickness and is characterized by nearly vertical 3- to 8-sided columns of basalt, with individual columns about 1 m in diameter and 7.5 m in length. The colonnade is usually less vesicular than the base. The entablature makes up about 70% of the flow thickness. This upper zone is characterized by small diameter (averaging less than 0.5 m) basalt columns which may develop into a fan-shape arrangement; hackly joints, with cross joints less consistently oriented and interconnected, may be rubbly and clinkery, and the upper part is vesicular. Following extrusion, flows cool rapidly, expelling gases and forming vesicles and cooling joints. These upper surfaces are typically broken by subsequent internal lava movement resulting in brecciated flow tops. The combination of the superposed flow base and vesicular upper part of the entablature is referred to as the interflow zone. Interflow zones generally make up 5 to 10% of the total flow thickness.

Groundwater occurrence and flow within layered sequences of basalt flows and intercalated sedimentary interbeds are complex. Such sequences typically consist of multiple zones of saturation with varying degrees of interconnection. Principle aquifers or water-bearing zones are associated with interflow zones between basalt flows. These interflow zones commonly have high to very high permeability and low storativity because of the open nature, but limited volume, of joints and fractures.

Furthermore, because of the generally impervious nature of the intervening tuffaceous sediments and dense basalt, stratigraphically adjacent interflow zones may be hydraulically isolated over large geographic areas. This physical and hydraulic separation is commonly reflected by differences in both piezometric levels and water quality between adjacent interflow aquifers (see Figure E.20).

Recharge occurs mainly along outcrops and through fractures that provide hydraulic communication to the surface. Interflow zones generally have the highest hydraulic conductivities and can form a series of superposed water-bearing zones. The colonnade and entablature are better con-

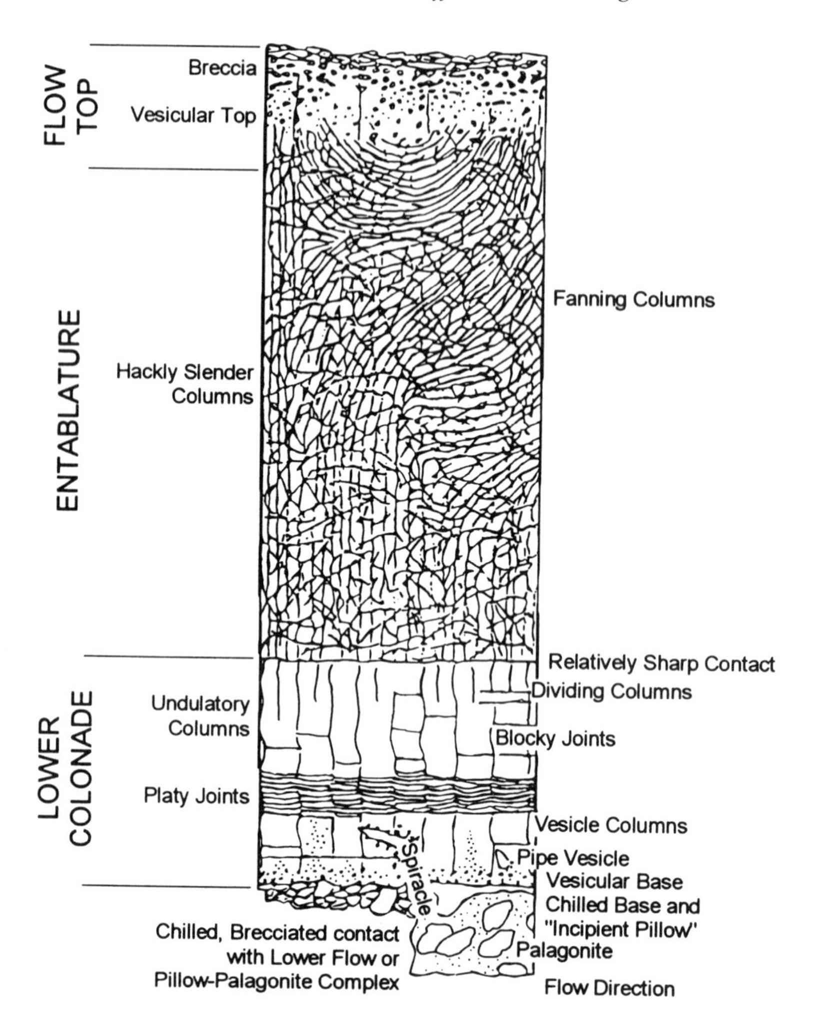

Figure E.19 Generalized schematic diagram showing intraflow structure of a basalt flow. (Modified after Swanson and Wright, 1978.)

nected vertically than horizontally, which allows for the movement of groundwater between interflow zones, although overall flow is three-dimensional. Multiple interflow zones can result in high total horizontal transmissivity. Position of the basalt flow within the regional flow system and varying hydraulic conductivities create further head differences with depth that can be very large in comparison to other sedimentary sequences. Horizontal hydraulic conductivities range from 0.65 up to 1600 or even 3000 m/day, whereas vertical hydraulic conductivities range from 10^{-8} to 10 m/day depending on the structural elements present (i.e., degree of fracturing joints, presence of sedimentary interbeds, etc.).

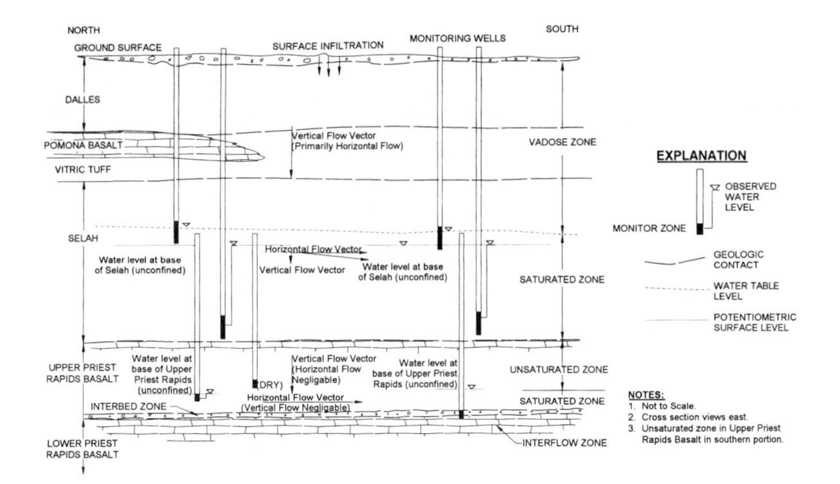

Figure E.20 Conceptual flow model showing zones of saturation flow vectors in relationship to observed water levels. (After Testa, 1988.)

Sedimentary interbeds are typically comprised of tuffaceous sediments of varying thickness, lateral extent, lithology, and degree of weathering. These interbeds usually impede groundwater movement in many areas. Groundwater flow within the more prominent interbeds is affected by the thickness and anisotropy of each hydrostratigraphic unit, and the position and continuity of each layer within the units. As with any layered media, the hydrostratigraphic unit with the lowest vertical hydraulic continuity is the controlling factor for groundwater flow in the vertical direction (normal to bedding). In the horizontal direction (parallel to bedding), groundwater flow is controlled by the hydrostratigraphic unit with the highest horizontal hydraulic continuity. Horizontal hydraulic conductivities based on pumping tests range from about 1×10^{-6} to 1×10^{-4} cm/s. Vertical hydraulic conductivities based on laboratory tests range from about 1×10^{-6} to 1×10^{-4} cm/s. Both methods showed a variance of 2 orders of magnitude.

E.5 Structural style and framework

Structural geologic elements that can play a significant role in subsurface environment-related issues include faults, fractures, joints, and shear zones. Faults can be important from a regional perspective in understanding their impact on the regional groundwater flow regime, and delineation and designation of major water-bearing strata. Faults are usually less important in most site-specific situations. Fractures, joints, and shear zones, however, canhave a significant role both regionally and locally in fluid flow and assessment of preferred migration pathways of dissolved MTBE in groundwater in consolidated and unconsolidated materials (see Table E.4). Regional

Table E.4 Sediment Types and Permeability for Different Sequences

Sequence	Sediment Type	Permeability
Marsh	Clay, silt, coal with a few aquifer sands of limited extent	Relatively low
Inner Bar	Sand with some clay/silt intercalations	Moderate to high
Outer Bar	Sand with many clay/silt intercalations	Moderate
Prodelta	Clay and silt	Low
Marine	Clay	Low

geologic processes that produce certain structural elements, notably fracture porosity, include faulting (seismicity), folding, uplift, erosional unloading of strata, and overpressing of strata.

Tectonic and possibly regional fractures result from surface forces (i.e., external to the body as in tectonic fractures); contractional and surface-related fractures result from body forces (i.e., internal to the forces). Contractional fractures are of varied origin resulting from desiccation, syneresis, thermal gradients, and mineral phase changes. Desiccation fractures develop in clay and silt-rich sediments upon a loss of water during subaerial drying. Such fractures are typically steeply dipping, wedge-shaped openings that form cuspate polygons of several nested sizes (Table E.5). Syneresis fractures result from a chemical process involving dewatering and volume reduction of clay, gel, or suspended colloidal material via tension or extension fractures. Associated fracture permeability tends to be isotropically distributed since developed fractures tend to be closely and regularly spaced. Thermal contractional fractures are caused by the cooling of hot rock, as with thermally induced columnar jointing in fine-grained igneous rocks (i.e., basalts). Mineral phase-change fracture systems are composed of extension and tension fractures related to a volume reduction due to a mineral phase change. Mineral phase changes are characterized by irregular geometry. Phase changes, such as calcite to dolomite or montmorillonite to illite, can result in about a 13% reduction in molar volume. Surface-related fractures develop during unloading, release of stored stress and strain, creation of free surfaces or unsupported boundaries, and general weathering. Unloading fractures or relief joints occur commonly during quarrying or excavation operations. Upon a one-directional release of load, the rock relaxes and spalls or fractures. Such fractures are irregular in shape and may follow topography. Free or unsupported surfaces (i.e., cliff faces, banks, etc.) can develop both extension andtensional fractures. These types of fractures are similar in morphology and orientation to unloading fractures. Weathering fractures are related to mechanical and chemical weathering processes such as freeze–thaw cycles,

Table E.5 Classification of Fractures

Fracture Type	Classification	Remarks
Experimental	Shear Extension Tensile	
Natural	Tectonic Regional Contractional Surface-Related	Due to surface forces Due to surface forces (?) Due to body force Due to body force

mineral alluation, diagenesis, small-scale collapse and subsidence, and mass-wasting processes.

E.5.1 Faults

Faults are regional structures that can serve as barriers, partial barriers, or conduits to groundwater flow and MTBE transport. The influence and effect of faults on fluid flow entrapment depend on the rock properties of strata that are juxtaposed and the attitude or orientation of the strata within their respective fault blocks. The influence of regional structural elements, notably faults, can have a profound effect on groundwater occurrence, regime, quality and usage, and delineation and designation of water-bearing zones of beneficial use.

The Newport-Inglewood Structural Zone in southern California exemplifies this important role (see Figure E.21). The structural zone is characterized by a northwesterly trending line of gentle topographic prominences extending about 40 miles.

This belt of domal hills and mesas, formed by the folding and faulting of a thick sequence of sedimentary rocks, is the surface expression of an active zone of deformation. An important aspect of this zone is the presence of certain fault planes that serve as effective barriers to the infiltration of seawater into the severely downdrawn groundwater aquifers of the coastal plain. These barriers also act as localized hydrogeologic barriers for freshwater on the inland side of the zone, reflected in the relatively higher water level elevations and enlarged effective groundwater aquifers.

The structural zone separates the Central groundwater basin to the northeast from the West Coast groundwater basin to the southwest. In the West Coast Basin area, at least four distinct water-bearing zones exist. In descending stratigraphic position, these zones are the shallow, unconfined Gaspur aquifer, the unconfined Gage aquifer of the upper Pleistocene Lakewood Formation, the semiconfined Lynwood Aquifer, and the confined Silverado aquifer of the lower Pleistocene San Pedro Formation.

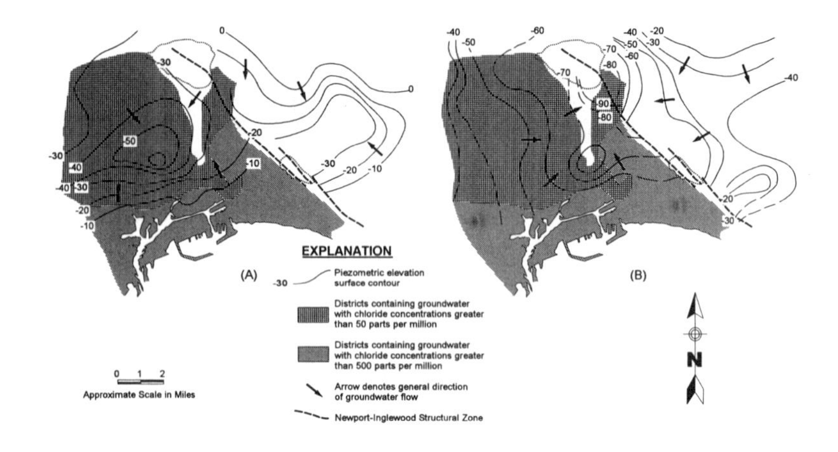

Figure E.21 Regional groundwater contour maps showing the Newport-Inglewood Structural Zone in relation to major water-bearing units, (After Testa, 1994.)

Groundwater conditions are strikingly different on opposing sides of the structural zone and are characterized by significant stratigraphic displacements and offsets, disparate flow directions, as much as 30 feet of differential head across the zone, and differences in overall water quality and usage. Shallow water-bearing zones situated in the area south of the structural zone have historically (since 1905) been recognized as being degraded beyond the point of being considered of beneficial use due to elevated sodium chlorides. Groundwater contamination (including MTBE) is also evident by the localized but extensive presence of light nonaqueous phase liquid (LNAPL) hydrocarbon pools and dissolved hydrocarbons due to the presence of 70 years of industrial development including numerous refineries, terminals, bulk liquid-storage tank farms, pipelines, and other industrial facilities on opposing sides of the structural zone.

The underlying Silverado aquifer has a long history of use, but has not been significantly impacted thus far by the poor groundwater quality conditions that have existed for decades in the shallower water-bearing zones where the Lynwood aquifer serves as a "guardian" aquifer. This suggests a minimal potential for future adverse impact of the prolific domestic-supply groundwater encountered at depths of 800 to 2600 ft below the crest of the structural zone. South of the structural zone, no direct communication exists between the historically degraded shallow and deeper water-bearing zones. The exception is in areas where intercommunication or leakage between water-bearing zones or heavy utilization of groundwater resources may (i.e., further to the northwest within the West Coast Basin). In contrast, north of the structural zone, shallow groundwater would be considered beneficial as a guardian aquifer due to the inferred potential for leakage into the deeper water-supply aquifers.

The beneficial use and cleanup standards thus are different north and south of the structural zone, with lower standards to the south. The overall environmental impact on groundwater resources, regardless of the ubiquitous presence of LNAPL pools and dissolved hydrocarbon plumes in certain areas relative to the structural zone, is minimal to nil. Within the structural zone, structures such as folds and faults are critical with respect to the effectiveness of the zone to act as a barrier to the inland movement of saltwater. An early continuous set of faults is aligned along the general crest of the structural zone, notably within the central reach from the Dominguez Gap to the Santa Ana Gap. The position, character, and continuity of these faults are fundamental to the discussion of groundwater occurrence, regime, quality, and usage. In addition, delineation and definition of aquifer interrelationships with a high degree of confidence are essential. The multifaceted impact of the structural zone is just one aspect of the level of understanding required prior to addressing certain regional groundwater issues. Another important issue is the assessment of which aquifers are potentially capable of beneficial use vs. those that have undergone historic degradation. Those faults that do act as barriers with respect to groundwater flow may, in fact, be one of several factors used in assigning a part of one aquifer to beneficial-use status as opposed to another. A second issue, based on the beneficial-use status, is the level of aquifer rehabilitation and restoration deemed necessary as part of the numerous aquifer remediation programs being conducted in the Los Angeles Coastal Plain. This example illustrates that, relative to aquifer remediation and rehabilitation efforts, cleanup strategies should not be stringent, nor should they be applied uniformly on a regional basis. Cleanup strategies should, however, take into account the complex nature of the hydrogeologic setting, and cleanup standards should be applied appropriately.

E.5.2 Fractured media

Fractured media in general can incorporate several structured elements including faults, joints, fractures, and shear zones.

These structural elements, as with faults, can serve as a barrier, partial barrier, or conduit to the migration of subsurface fluids. Most fractured systems consist of rock or sediment blocks bounded by discrete discontinuities. The aperture can be open, deformed, closed, or a combination thereof. The primary factors to consider in the migration of subsurface fluids within fractured media are fracture density, orientation, effective aperture width, and nature of the rock matrix. Fracture networks are complex three-dimensional systems. The analysis of fluid flow through a fractured media is difficult since the only means of evaluating hydraulic parameters is by means of hydraulic tests. The conduct of such tests requires that the geometric pattern or degree of fracturing formed by the structural elements (i.e., fractures) be known. Fracture density (or the number of fractures per unit volume of rock) and orientation are most important in assessing the degree of interconnection of fracture sets. Fracture spacing is influenced by mechanical behavior (i.e., interactions of intrinsic properties). Intrinsic and environmental properties include load-bearing framework, grain size, porosity, permeability, thickness, and previously existing mechanical discontinuities. Environmental properties of importance include net overburden, temperature, time (strain rate), differential stress, and pore fluid composition. Fracturing can also develop under conditions of excessive fluid pressures. Clay-rich soils and rocks, for example, are commonly used as an effective hydraulic seal. However, the integrity of this seal can be jeopardized if excessive fluid pressures are induced, resulting in hydraulic fracturing. Hydraulic fracturing in clays is a common feature in nature at hydrostatic pressures ranging from 10 KPa up to several MPa. Although hydraulic fracturing can significantly decrease the overall permeability of the clay, the fractures are likely to heal in later phases due to the swelling pressure of the clay.

Several techniques have been used to attempt to characterize fracture networks. These techniques have included field mapping (i.e., outcrop mapping, lineation analysis, etc.), coring, aquifer testing, tracer tests, borehole flowmeters, statistical methods, geophysical approaches, and geochemical techniques to evaluate potential mixing. Vertical parallel fractures are by far the most difficult to characterize for fluid flow analysis due to the likelihood of their being missed during any drilling program. This becomes increasingly important because certain constituents such as solvents and chlorinated hydrocarbons, that are denser than water, are likely to migrate vertically downward through the preferred pathways, and may even increase the permeability within these zones.

Within a single set of measured units of the same lithologic characteristics, a linear relationship is assumed between bed thickness and fracture spacing. A typical core will intercept only some of the fractures. In viewing the schematic block diagram of a well bore through fractured strata of varying thicknesses, the core drilled in the upper and lower beds will intersect fractures, but the cores drilled within the two central beds do not encounter any fractures. Closer fracture spacing is, however, evident in the two upper thinner beds.

The probability of intercepting a vertical fracture in a given bed is given by

$$P = \frac{D}{S} = \frac{DI}{T(\text{average})} \qquad \text{(eq. 3)}$$

where P = probability, D = core diameter, S = distance between fractures, T(average) is average thickness, and I is a fracture index given by:

$$I = \frac{Ti}{Si} \qquad \text{(eq. 4)}$$

where the subscript i refers to the properties of the bed. In other words, T_i is thickness of the i-th bed and S_i is fracture spacing in the i-th bed. A fracture index must also be determined independently for each set of fractures in the core, and must be normal to the bedding.

Based on probability, a core has a finite chance of intersecting a vertical fracture in a bed of a given thickness depending on core diameter, bed thickness, and the value of the fracture index (I). Thus, a sparsely fractured region has a small value of I (i.e., large spacing between parallel fractures), and thicker beds have a larger spacing between fractures for a given I.

E.6 Seismicity

Earthquakes can cause significant changes in water quality and water levels, ultimately enhancing permeability. During the Loma Prieta, California earthquake of October 17, 1989 (magnitude 7.1), ionic concentrations and the calcite salination index of streamwater increased, streamflow and solute concentrations decreased significantly from within 15 minutes to several months following the earthquake, and groundwater levels in the highland parts of the basins were locally lowered by as much as 21 m within weeks to months after the earthquake. The spatial and temporal character of the hydrologic response sequence increased rock permeability and temporarily enhanced groundwater flow rates in the region as a result of the earthquake.

Endnotes and references

This appendix was modified and updated from Testa, Stephen M., 1994, Chapter 3, Geologic Principles from *Geologic Aspects of Hazardous Waste Management*, pp. 55–99, Lewis Publishers, Boca Raton, FL.

Allan, U.S., Model for Hydrocarbon Migration and Entrapment Within Faulted Structures, *American Association of Petroleum Geologists Bulletin*, v. 73, no. 7, pp. 803–811, 1989.

Allen, J.R.L., *Physical Processes of Sedimentation*, Allen & Urwin, London, 248 p., 1997.

Allen, J.R.L., Studies in Fluviatile Sedimentation, An Exploratory Quantitative Model for the Architecture of Avulsion-Controlled Alluvial Suites: *Sedimentary Geology*, 21, pp. 129–147, 1978.

Anderson, M.P., Hydrogeologic Facies Models to Delineate Large-Scale Spatial Trends in Glacial and Glaciofluvial Sediments, *Geological Society of America Bulletin*, v. 101, pp. 501–511, 1989.

Aquilera, R., Determination of Subsurface Distance between Vertical Parallel Natural Fractures based on Core Data: *American Association of Petroleum Geologists Bulletin*, v. 72, no. 7, pp. 845–851, 1988.

Back, W., Rosenshein, J.S., and Seaber, P.R., *Hydrogeology, The Geology of North America:* Vol. 0-2, Geological Society of America, Boulder, CO, 524 p., 1988.

Barker, J.F., Barbash, J.E. and Labonte, M., Groundwater Contamination at a Landfill Sited on Fractured Carbonate and Shale: *Journal of Contaminant Hydrology,* v. 3, pp. 1–25, 1988.

Beck, B.F., *Engineering and Environmental Impacts of Sinkholes and Karst,* A.A. Balkema, Rotterdam, The Netherlands, 384 p., 1989

Berg, R.C. and Kempton, J.P., Potential for Contamination of Shallow Aquifers from Land Burial of Municipal Wastes, Illinois State Geological Survey Map, Scale I :500,000, 1984.

Berg, R.C., Kempton, J.P., and Cartwright, K., Potential for Contamination of Shallow Aquifers in Illinois: Illinois State Geological Survey Circular 532, 30 p., 1984.

Caswell, B., Time-of-Travel in Glacial Aquifers, *Water Well Journal,* March, pp. 48–51, 1988a.

Caswell, B., Esker Aquifers, *Water Well Journal,* July, pp. 36, 37, 1988b.

Cayeux, L., *Carbonate Rocks:* (translated by A.T. Carozzi), Hafner Publishing, Darien, CT, 506 p., 1970.

Cehrs, D., Depositional Control of Aquifer Characteristics in Alluvial Fans, Fresno County, California: *Geological Society of America Bulletin,* v. 90, no. 8, Part I, pp. 709–711, Part II, pp. 1282- 1309, 1979.

Chandler, M.A., Kocurek, G., Goggin, D.J., and Lake, L.W., Effects of Stratigraphic Heterogeneity on Permeability in Eolian Sandstone Sequence, Page Sandstone, Northern Arizona, *American Association of Petroleum Geologists Bulletin,* v. 73, no. 5, pp. 658–668, 1989.

Choquette, P.W. and Pray, L.C., Geologic Nomenclature and Classification of Porosity in Sedimentary Carbonates, *American Association of Petroleum Geologists Bulletin,* v. 54, pp. 207–250, 1970.

Coleman, J.M. and Wright, L.D., Modern River Deltas: Variability of Processes and Sand Bodies, in *Deltas: Models for Exploration* (edited by Broussard, M.), Houston Geological Society, 1975.

Currie, J.B., Significant Geologic Processes in Development of Fracture Porosity, *American Association of Petroleum Geologists Bulletin,* v. 61, no. 7, pp. 1086–1089, 1977.

Davidson, K.S., Geologic Controls on Oil Migration Within a Coastal Limestone Formation, in Proceedings of the Association of Groundwater Scientists and Engineers and the American Petroleum Institute Conference on Petroleum Hydrocarbons and Organic Chemicals in Groundwater: Prevention, Detection and Restoration, Houston, pp. 233–251, 1988.

DeGear, J., Some Hydrogeological Aspects on Aquifers, especially Eskers: in *Groundwater Problems,* Erickson, E., Gustafsson, Y. and Nilson, K. Eds., Pergamon Press, New York, pp. 358–364, 1968.

Driscoll, Fletcher G, *Groundwater and Wells,* 2nd edition, Johnson Filtration Company, Inc., St. Paul, Minnesota, p. 31, 1986.

Ebanks, W.J., Jr., Geology in Enhanced Oil Recovery, in *Reservoir Sedimentology,* Tillman, R.W. and Weber, K.J., Society of Economic Paleontologists and Mineralogists, Special Publication no. 40, pp. 1–14, 1987.

Fogg, G.E., Emergence of Geologic and Stochastic Approaches for Characterization of Heterogeneous Aquifers, in Proceedings of the U.S. Environmental Protection Agency, Robert S. Kerr Environmental Research Laboratory Conference on New Field Techniques for Quantifying the Physical and Chemical Properties of Heterogeneous Aquifers, March 20–23, 1989, Dallas, pp. 1–17, 1989.

Folk, R.L., *Petrology of Sedimentary Rocks,* Hemphill Publishing, Austin, TX, 159 p., 1974.

Freeze, R.A. and Cherry, J.A., *Groundwater*, Prentice Hall, New York, pp. 145–166, 1979.

Fryberger, S.G., Stratigraphic Traps for Petroleum in Wind-Laid Rocks, *American Association of Petroleum Geologists Bulletin,* v. 70, no. 12, pp. 1765–1776, 1986.

Gregory, K.J., Ed., *Background to Paleohydrology - A Perspective:* John Wiley & Sons, New York, 486 p., 1983.

Grim, R.E., *Clay Mineralogy,* McGraw-Hill, New York, 596 p., 1968.

Haberfeld, J.L., Hydrogeology of Effluent Disposal Zones, Floridan Aquifer, South Florida: *Groundwater,* v. 29, no. 2, pp. 186–190, 1991.

Halbouty, M.T., The Deliberate Search for the Subtle Trap, American Association of Petroleum Geologists Memoir no. 32, Tulsa, OK, 351 p., 1982

Harman, Jr., H.D., Detailed Stratigraphic and Structural Control, The Keys to Complete and Successful Geophysical Surveys of Hazardous Waste Sites, in Proceedings of the Hazardous Materials Control Research Institute Conference on Hazardous Wastes and Hazardous Materials, Atlanta, pp. 19–21, 1986.

Harris, D.G., The Role of Geology in Reservoir Simulation Studies, *Journal of Petroleum Technology,* May, pp. 625–632, 1975.

Hitchon, B., Bachu, S., and Sauveplane, C.M., Eds., Hydrogeology of Sedimentary Basins - Application to Exploration and Exploitation, in Proceedings of the National Water Well Association Third Canadian/American Conference on Hydrogeology, June 22–26, 1986, Banff, Alberta, Canada, 275 p., 1986.

Hitchon, B. and Bachu, S. Eds., Fluid Flow, Heat Transfer and Mass Transport in Fractured Rocks, in Proceedings of the National Water Well Association Fourth Canadian/American Conference on Hydrogeology, Banff, Alberta, 283 p., 1988.

Jardine, D. and Wilshart, J.W., Carbonate Reservoir Description, in *Reservoir Sedimentology,* R.W. Tillman and K. J. Weber, Eds., Society of Economic Paleontologists and Mineralogists, Special Publication no. 40, p. 129, 1987.

Keller, B., Hoylmass, E., and Chadbourne, J., Fault Controlled Hydrology at a Waste Pile: *Groundwater Monitoring Review,* Spring, pp. 60–63, 1987.

Kerans, C., Karst-controlled Reservoir Heterogeneity in Ellenburger Group Carbonates of West Texas, *American Association of Petroleum Geologists Bulletin,* v. 72, no. 10, pp. 1160–1183, 1988.

Larkin, R.G. and Sharp, J.M., Jr., On the Relationship Between River-Basin Geomorphology, Aquifer Hydraulics, and Groundwater Flow Direction in Alluvial Aquifers: *Geological Society of America Bulletin,* v. 104, pp. 1608–1620, 1992.

Lindholm, G.F. and Vaccaro, J.J., Region 2, Columbia Lava Plateau, in *The Geology of North America, v. 0-2, Hydrogeology,* Geological Society of America, Boulder, CO, pp. 37–50, 1988.

Magner, J.A., Book, P.R., and Alexander, Jr., E.C., A Waste Treatment/Disposal Site Evaluation Process for Areas Underlain by Carbonate Aquifers, *Groundwater Monitoring Review,* Spring, pp. 117–121, 1986.

Mancini, E.A., Mink, R.M., Bearden, B.L., and Wilkerson, R.P., Norphlet Formation (Upper Jurassic) of Southwestern and Offshore Alabama: Environments of Deposition and Petroleum Geology, *American Association of Petroleum Geologists Bulletin,* v. 69, no. 6, pp. 881–898, 1985.

Miall, A.D., Reservoir Heterogeneities in Fluvial Sandstones, Lessons from Outcrop Studies; *American Association of Petroleum Geologists Bulletin,* v. 72, no. 6, pp. 682–697, 1988.

Millot, G., *The Geology of Clays,* Springer-Verlag, New York, 429 p., 1970.

Moran, M.S., Aquifer Occurrence in the Fort Payne Formation, *Groundwater,* v. 18, no. 2, pp. 152–158, 1980.

Nan, W. and Lerche, I., A Method for Estimating Subsurface Fracture Density in Core, *American Association of Petroleum Geologists Bulletin,* v. 68, no. 5, pp. 637–648, 1984.

Nelson, R.A., *American Association of Petroleum Geologists Bulletin,* v. 63, no. 12, pp. 2214–2232, 1979.

Nilsen, T.H., Alluvial Fan Deposits, in *Sandstone Depositional Environments,* P.A. Scholle and D. Spearing, Eds., American Association of Petroleum Geologists, Memoir no. 31, Tulsa, OK, pp. 49–86, 1982.

Osciensky, J.L., Winter, G.V., and Williams, R.E., Monitoring and Mathematical modeling of Contaminated Groundwater Plumes in Paleofluvial Environments for Regulatory Purposes, in Proceedings of the National Water Well Association Third National Symposium on Aquifer Restoration and Groundwater Monitoring, pp. 355–364, 1983.

Pettijohn, F.J., Potter, P.E., and Siever, R., *Sand and Sandstone,* Springer-Verlag, New York, 618 p., 1973.

Poeter, E. and Gaylord, D.R., Influence of Aquifer Heterogeneity on MTBE Transport at the Hanford Site: *Groundwater,* v. 28, no. 6, pp. 900–909, 1990.

Potter, P.E. and Pettijohn, F.J., *Paleocurrents and Basin Analysis,* Springer-Verlag, New York, 296 p., 1977.

Pryor, W.A., Permeability-Porosity Patterns and Variations in Some Holocene Sand Bodies, *American Association of Petroleum Geologists Bulletin,* v. 57, no. 1, pp. 162–189, 1973.

Reading, H.G., Ed., *Sedimentary Environments and Facies,* Blackwell Scientific Publications, Oxford, 557 p., 1978.

Reineck, H.E. and Singh, I.B., *Depositional Sedimentary Environments With Reference to Terrigenous Clastics,* Springer-Verlag, New York, 439 p., 1975.

Rojstaczer, S. and Wolf, S., Permeability Changes Associated with Large Earthquakes: An Example from Loma Prieta, Califomia, *Geology,* v. 20, pp. 211-214, 1992.

Scherer, M., Parameters Influencing Porosity in Sandstones, A Model for Sandstone Porosity Prediction, *American Association of Petroleum Geologists Bulletin,* v. 71, No. 5, pp. 485-491, 1987.

Scholle, P.A., *A Color Illustrated Guide to Carbonate Rock Constituents, Textures, Cements and Porosities of Sandstones and Associated Rocks,* American Association of Petroleum Geologists Memoir no. 27, Tulsa, OK, 241 p., 1979.

Scholle, P.A., *A Color Illustrated Guide to Constituents, Textures, Cements and Porosities of Sandstones and Associated Rocks,* American Association of Petroleum Geologists Memoir no. 28, Tulsa, OK, 201 p., 1979.

Scholle, P.A. and Spearing, D., *Sandstone Depositional Environments,* American Association of Petroleum Geologists Memoir no. 31, Tulsa, OK, 410 p., 1982.

Scholle, P.A., Bebout, D.G., and Moore, C.H., *Carbonate Depositional Environments,* American Association of Petroleum Geologists Memoir No. 33, Tulsa, OK, 708 p., 1983.

Sciacca, J., Essential Applications of Depositional Analysis and Interpretations in Hydrogeologic Assessments of Contaminated Sites, in Proceedings of the Hazardous Materials Control Research Institute HWHM/HMC-South '91 Conference, Houston, pp. 294–298, 1991.

Sciacca, J., Carlton, C., and Gios, F., Effective Application of Stratigraphic Borings and Analysis in Hydrogeologic Investigations, in Proceedings of the Hazardous Materials Control Research Institute HWHM/HMC-South '91 Conference, Houston, pp. 69–74, 1991.

Schmelling, S.G. and Ross, R.R., MTBE Transport in Fractured Media, Models for Decision Makers, U.S. Environmental Protection Agency EPA/5401489/004, 8 p., 1989.

Selley, R.C., *An Introduction to Sedimentology,* Academic Press, London, 408 p., 1976.

Selley, R.C., *Ancient Sedimentary Environments,* Chapman and Hall, London, 287 p., 1978.

Sharp, J.M., Hydrogeologic Characteristics of Shallow Glacial Drift Aquifers in Dissected Till Plains (North-Central Missouri), *Groundwater,* v. 22, no. 6, pp. 683–689, 1984.

Sneider, R.M., Tinker, C.N., and Meckel, L.D., Deltaic Environment Reservoir Types and Their Characteristics, *Journal of Petroleum Technology,* pp. 1538- 1546, 1978.

Soller, D.R. and Berg, R.C., A Model for the Assessment of Aquifer Contamination Potential Based on Regional Geologic Framework, *Environmental Geology and Water Science,* v. 19, no. 3, pp. 205-213, 1992.

Spearing, D.R., Alluvial Fan Deposits - Summary Sheet of Sedimentary Deposits: Sheet 1, Geological Society of America, Boulder, CO.,1974.

Stephenson, D.A., Fleming, A.H., and Mickelson, D.M., The Hydrogeology of Glacial Deposits, in *Hydrogeology,* Back, W., Rosenshein, J.S., and Seaber, P.R. Eds., Geological Society of America Decade of North American Geology, Boulder, CO, v. 0-2, pp. 301–304, 1989.

Testa, S.M., Regional Hydrogeologic Setting and its Role in Developing Aquifer Remediation Strategies, in Proceedings of the Geological Society of America Annual Meeting, Abstracts with Programs, November 6–9, 1989, St. Louis, pp. A96, 1989.

Testa, S.M., Site Characterization and Monitoring Well Network Design, Columbia Plateau Physiographic Province, Arlington, North-Central Oregon, Geological Society of America Abstract, 325 p., 1991.

Testa, S.M., Henry, E.C., and Hayes, D., Impact of the Newport-Inglewood Structural Zone on Hydrogeologic Mitigation Efforts - Los Angeles Basin, California, in Proceedings of the Association of Groundwater Scientists and Engineers FOCUS Conference on Southwestern Groundwater Issues; Albuquerque, NM, pp. 181–203, 1988.

Testa, S.M. and Winegardner, D.L., *Restoration of Contaminated Aquifers, Petroleum Hydrocarbons and Organic Compounds,* CRC Press, Boca Raton, FL, 446 pp., 2000.

Tyler, N., Gholston, J.C. and Ambrose, W.A., Oil Recovery in a Low Permeability, Wave-Dominated, Cretaceous, Deltaic Reservoir, Big Wells (San Miguel) Field, South Texas, *American Association of Petroleum Geologists Bulletin,* v. 71, no. 10, pp. 1171–1195, 1987.

Wang, C.P. and Testa, S.M., Groundwater Flow Regime Characterization, Columbia Plateau Physiographic Province, Arlington, North-Central Oregon, in Proceedings of the Association of Groundwater Scientists and Engineers Conference on New Field Techniques for Quantifying the Physical and Chemical Properties of Heterogeneous Aquifers, pp. 265–291, 1989.

Weber, K.J., Influence on Fluid Flow on Common Sedimentary Structures in Sand Bodies, Society of Petroleum Engineers, SPE Paper 9247, 55th Annual Meeting, pp. 1–7, 1980.

Weber, K.J., Influence of Common Sedimentary Structures on Fluid Flow in Reservoir Models, *Journal of Petroleum Technology*, March, pp. 665–672, 1982.

Yaniga, P.M. and Demko, D.J., Hydrocarbon Contamination of Carbonate Aquifers, Assessment and Abatement, in Proceedings of the National Water Well Association Third National Symposium on Aquifer Restoration and Groundwater Monitoring, pp. 60–65, 1983.

Appendix F

MTBE: subsurface investigation and cleanup

F.1 Introduction

Since methyl *tertiary*-butyl ether (MTBE) moves faster than the associated BTEX plume, and because MTBE does not naturally degrade in the environment, the problem of discovering MTBE contamination at properties that never contained underground gasoline storage tanks will likely continue for decades. Proper MTBE sampling and assessment protocol is appropriate for environmental due diligence during property transfers.

F.2 Subsurface environmental evaluation: phases I through IV

Subsurface (soil and groundwater) environmental evaluations of potential spills or leaks of MTBE and other hazardous substances generally consist of four phases (Phases I through IV). The generalized four phases reflect the pragmatic process required to take a potentially MTBE-impacted property from the initial, assessment phase (Phase I) through to the final monitoring and site closure phase (Phase IV). If the property in question was known to have previously contained an underground gasoline storage tank with MTBE, then the evaluation may start with a Phase II assessment or even a Phase III evaluation, in the case of a known MTBE release. If MTBE is discovered in a well, then a Phase I assessment may be performed to evaluate adjacent properties in the area local to the contaminated site for the presence of a former or current underground tank that might have contained MTBE.

F.2.1. Phase I: environmental assessment

Phase I environmental assessment consists of the application of a set of noninvasive techniques to acquire information through site inspection and to obtain data supplied by others through appropriate research. Phase I assessments involve a site inspection, the determination of the history of the subject property as well as nearby properties, and review of data obtained from environmental lists, building permit databases, regulators, owners, tenants, and others. Techniques include interviews of persons knowledgeable about the site, review of historical aerial photography, and examination of published and unpublished maps. Phase I may also coincide with the transfer of property ownership, and may be part of environmental due diligence. As part of Phase I, an internet-accessible database such as the California GeoTracker, a geographic information system with an interface to the Geographic Environmental Information Management System (GEIMS) provides a wealth of detailed information on many sites, including data regarding underground gasoline storage tanks, MTBE releases, and public drinking water supplies.

F.2.2 Phase II: subsurface investigation

Phase II subsurface investigation is an invasive method of site evaluation and is designed to evaluate the geologic and hydrogeologic conditions of the site in question by collecting soil, soil vapor, and groundwater samples. A variety of different techniques and equipment are currently available for assessing the subsurface at this level of investigation. Environmental subsurface investigation tools range in size, cost, and operating complexity, from hand augers and hand-operated drive samplers, to direct push technology (DPT) rigs, to rotary rigs. Borings are drilled for geologic, hydrogeologic, and lithologic characterization. Samples are collected for field screening and physical testing. Selected soil, vapor, or groundwater samples are submitted to a certified laboratory for chemical testing.

Phase II of subsurface contamination investigation at a site can take place in several stages. Completion of Phase II investigation, or investigations, is reached when the vertical and lateral extent of the soil and groundwater contamination has been fully characterized. The first Phase II investigation may be of limited extent wherein only three or four borings may be drilled; groundwater monitoring wells may also be installed. Several Phase II subsurface investigations may be required in order to fully characterize the extent of the MTBE soil and groundwater contamination. The level of detail of a Phase II investigation and characterization program is generally related to a variety of factors, including contamination source, subsurface pathway and receptor conditions.[1]

The goals of groundwater monitoring are described as follows:[2]

- To define the vertical and aerial extent of groundwater contamination.
- To monitor target chemical concentrations over time.
- To provide a measure for detecting the contaminant plume front, unexpected changes in plume size or direction of flow.
- To determine the extent of interaquifer movement of contaminants.
- To determine aquifer characteristics such as permeability, transmissivity, etc.
- To estimate the rate of contaminant plume movement.

- To develop a database for designing remedial measures.
- To determine the effects of the remedial measures.
- To assist in performing the remedial work (provide hydraulic control and contaminant removal).
- To provide databases for groundwater modeling.
- To evaluate the aquifer during regular sampling events over the yearly hydrologic cycle.

F.2.3 Phase III: corrective action

Phase III, corrective action, is the remediation portion of a subsurface contamination investigation and cleanup project. This phase involves designing and implementing a corrective action plan to effectively remediate contaminated soil and/or groundwater. A corrective action plan evaluating remedial options and their feasibility is typically submitted to the lead regulator in charge of oversight of site cleanup operations prior to commencement of field work to clean up the site. Remediation of MTBE may include engineered bioremediation (in situ), chemical oxidation (in situ), soil excavation and removal to a landfill, or other technologies that are first presented in the corrective action plan.

F.2.4 Phase IV: site monitoring and closure

After remediation has been achieved at the site, the regulatory agencies with authority over the site generally require at least 1 year of quarterly groundwater monitoring to verify that the majority of the source of the contamination had been removed. Phase IV includes this sampling as well as the proper abandonment of existing groundwater monitoring wells after site closure has been obtained. If the site is a low-risk MTBE site, wherein (1) the sampling data suggest that MTBE has been released in low to moderate amounts in the soil, (2) the groundwater has been impacted in a limited vertical and lateral extent, (3) the estimated rate of groundwater recharge is relatively low, and (4) groundwater is not used as a source of drinking water, a risk-based corrective action (RBCA) strategy is one effective method of establishing the potential pathways and risks of exposure to justify site closure.

F.3 Drilling and MTBE sampling techniques

A variety of different drilling techniques have been developed and are used in the environmental field. Drilling and sampling techniques for sites with MTBE contamination are the same as those applied at any typical gasoline release contamination site, whether or not MTBE is present at the site. The following section describes drilling and sampling techniques applied at typical gasoline release cleanup sites.

The factors that determine the drilling technique to be used at a site with subsurface contamination relate to timing and cost of the cleanup project, sediment type (consolidated rock or unconsolidated soils), sample type (whether the area from which the sample is taken is undisturbed or disturbed), and sample integrity. Unconsolidated deposits are drilled primarily using Direct Push Technology rigs. Cone penetrometer testing (a subset of Direct Push Technology) drilling can also be used. For deeper drilling projects in difficult drilling conditions, hollow-stem augering techniques could be used. In some areas of the country, even cable-tool drilling techniques are still used for environmental sampling and well installation. For consolidated or semi-consolidated deposits, continuous wire-line or conventional rock coring techniques are commonly used.

Other drilling techniques used in the environmental field include the use of labor-intensive hand augers for shallow depths, and machine-operated solid-flight augers. Both of these techniques can result in sample retrieval problems, because samples can only be collected if the entire auger is removed prior to sampling.

A summary of the pros and cons of various groundwater sampling tools is presented in Table F.1.

F.4 Overview of drilling methods

F.4.1. Direct push technology sampling methods

The following presents an overview of drilling methods for environmental sampling. Table F.1 is a summary of the pros and cons of different tools used for environmental sampling. Table F.2 describes various sampling monitoring tools.

Direct Push Technology (DPT; also called Drive Point Sampling) is a quicker and less costly alternative sampling method than conventional rotary drilling for collecting soil, vapor, and water samples for environmental projects (Figure F.1). DPT equipment allows fewer permanent monitoring wells, multiple depth sampling programs, elimination or minimization of drilling-derived wastes, the ability to perform on-site chemical analysis, and minimal exposure of workers to potentially hazardous soil cuttings.

DPT sampling relies on dry impact methods to push or hammer boring and sampling tools into the subsurface for environmental assessments. This technology does not require the use of hazardous chemicals, drilling fluids, or water during operation. A typical, conventionally augered borehole drilled to 60 ft would generate approximately 6 drums of soil cuttings (waste soil generated as a byproduct of drilling). DPT equipment produces soil samples but generally does not produce soil cuttings.

DPT works well with a variety of lithologies, including clays, silts, sands, and gravels; however, this technology is not designed to penetrate or sample bedrock. DPT equipment has been used successfully in limited access areas

Table F.1 Groundwater Sampling Tools: Pros and Cons

Sampling Tool	Pros	Cons
BAT® Enviroprobe	1) Very accurate (uses pre-evacuated vials) 2) Durable 3) Produces very clear samples 4) Fill detection capability 5) Can be used for permeability testing in vadose and saturated zones 6) Can be used as a soil gas sampler	1) Very expensive 2) Very limited sample volume 3) Long time to deploy and retrieve 4) Long groundwater sample recovery time in sandy silts, silts, and clays 5) Very complex for first time use 6) Must be at least 5 ft below first water to collect groundwater samples
MaxiSimulProbe®	1) Very accurate (uses pressurized water canister) 2) Large sample volume 3) Collects core simultaneously with water or vapor 4) Very durable 5) Can be wire lined with down hole hammer for rapid deployment and retrieval 6) Excellent penetration capability in dense sediments 7) Fill detection capability	1) Expensive 2) Moderately complex for first time use 3) Long groundwater sample recovery time in sandy silts, silts and clays 4) Can only be used with conventional drilling machines (HSA, MR, DWP, ARCH) 5) Must be at least 5 ft below first water to collect groundwater samples
MiniSimulProbe®	1) Very accurate (uses pressurized water canister) 2) Large sample volume 3) Collects core simultaneously with water or vapor 4) Durable 5) Can be wire lined with down hole hammer for rapid deployment and retrieval 6) Can be used with CPT as well as conventional drilling machines (HAS, MR, DWP, ARCH) 7) Good penetration capability in dense sediments 8) Low turbidity 9) Fill detection capability	1) Expensive 2) Very complex for first time use 3) Long groundwater sample recovery time in sand silts, silts, and clays as well as silty sands 4) Must be at least 5 ft below first water to collect groundwater samples
H2-Vape Probe®	1) Very accurate (uses pressurized water canister) 2) Large sample volume 3) Long screen length 4) Very durable 5) Can be wire lined with down hole hammer for rapid deployment and retrieval 6) Fairly rapid groundwater sample recovery time 7) Low turbidity 8) Collect soil gas, then push downward for groundwater sample	1) Tool is expensive 2) Consumables are moderately expensive 3) Must be at least 5ft below first water to collect groundwater samples

Table F.1 (continued)

Sampling Tool	Pros	Cons
HydroPunch® (HPI) and HydroPunch® (HPII) Hydrocarbon Mode	1) Excellent first water samplers due to long screen length and bailer access into screen 2) Very simple 3) Low cost per sample 4) Widely used and accepted in industry 5) Can be used with all types of drilling rigs including CPT and vibratory direct push	1) VOC loss due to bailer use Limited for vertical profiling projects due to cross contamination and/or dilution from leaky rod joints, Generally produces turbid samples
HPI and HP II Groundwater Mode	None 1) Recoverable drive cone 2) Can be used with direct push rigs 3) Can hold 1 liter water sample	1) Canister must be completely full to maintain sample equilibrium and to prevent cross contamination from hole fluids Fairly complex for first time use Average to below average durability, Tool is expensive, Consumables are moderately expensive, Generally produces fairly turbid water samples, Can never really be sure how much of groundwater sample is from formation and how much is from bore hole.
Cone Sipper®	1) Very accurate (uses inert gas drive system) 2) Durable 3) Continuous profiler in vadose and saturated zone (can collect continuous soil gas and ground water samples in one push) 4) Can be stacked onto an electric friction cone (CPT cone) and used to collect groundwater samples from permeable zones immediately after electric friction cone identifies soil and permeability type 5) No cuttings 6) Can be used with back grouting CPT cone 7) Low sample turbidity	1) Can only be used with CPT 2) Very expensive 3) Moderately complex for first time use 4) Screen can be easily plugged when tool passes through fine grain sediment 5) Typically limited to coarse sands with little to no silt content
Waterloo Profiler®	1) Can be very accurate when used with peristaltic pump provided there are limited dissolved gases in sample 2) Durable 3) Continuous profiler (saturated zone only 4) Can be used with vibratory direct push as well as CPT 5) No cuttings 6) Is available with back grouting feature 7) Low sample turbidity	1) Expensive when take into account all of the required accessories, Moderately complex for first time use, Small fluid entry ports, Screens can be easily plugged when tool passes through fine grained sediments, Typically limited to coarse sands with little to no silt content, Loses accuracy when used with bailer or when using peristaltic pump with high dissolved gas concentration.
Geoprobe® Groundwater Profiler	1) Can be very accurate when used with peristaltic pump provided there are limited dissolved gases in sample, Durable, Continuous profiler (saturated zone only), Can be used with vibratory direct push as well as CPT, Long screen length to expedite groundwater sample recovery, No cuttings, Available with back grouting capability, Low sample turbidity	1) Expensive when take into account all of the required accessories, Moderately complex for first time use, Screens can be easily plugged when tool passes through fine grained sediments,Typically limited to coarse sands with little to no silt content, Loses accuracy when used with bailer or when using peristaltic pump with high dissolved gas concentration, Restricted to use with vibratory direct push

(After Heller and Jacobs, used by permission.)

such as in basements of buildings, under canopies, and inside buildings. DPT sampling has also been used successfully in sensitive environments such as wetlands, tundra, lagoon, and bay settings.

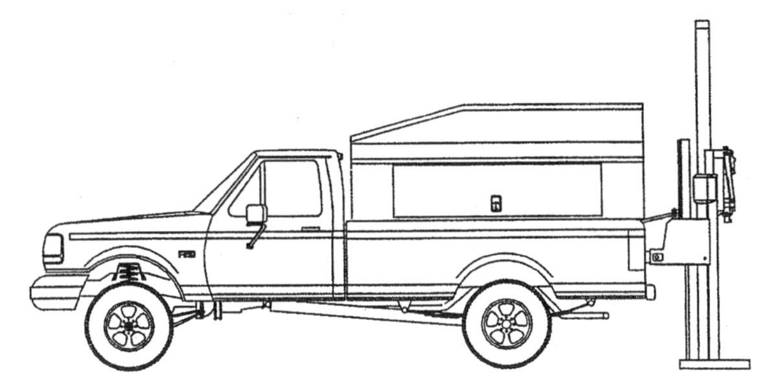

Figure F.1 Probe or DPT rig. (From Geoprobe®, 1999. With permission.)

F.4.2 DPT probe rigs and CPT rigs

DPT is a technique that extracts soil, vapor, or water samples through the use of samplers that are driven into the subsurface without the rotary action associated with hollow-stem auger rigs. DPT rigs, including cone penetrometer testing (CPT) rigs, use the static weight of the vehicle to push the sampling rods into the ground. CPT rigs use a 20-ton truck and are capable of sampling to depths of 250 ft. CPT rigs, originally developed for use in the geotechnical field, typically push from the center of the truck.

Small, highly maneuverable direct push probes were developed in the late 1980s. The probes were placed on pick-up trucks and vans. Probe rigs generally push the rods from the back of the truck. A percussion hammer can be added to these probe units to enhance the depth of sampling. These smaller probes have lowered the cost of DPT sampling projects that drill to depths of 60 ft or less. Truck-mounted DPT probe rigs are typically hydraulically powered. The percussion/probing equipment pushes rods connected to small-diameter (0.8- to 3.0-in.) samplers.

F.4.2.1 DPT probe soil sampling

The soil samples extracted through the use of DPT probes are commonly collected in 2- to 5-ft long clear plastic (polyethylene or butyrate) liners contained within an outer sampler. The plastic liners are easily cut with a knife and are transparent for easy lithologic characterization. Brass, aluminum, stainless steel, or Teflon® liners are also available, depending on the sampler. After removal from the sampler, the soil liner containing the sample is immediately capped on both ends with Teflon® tape, trimmed, and then capped with plastic caps. The samples are then labeled and placed in individual transparent, hermetically sealed sampling bags. The samples are then placed in an appropriate refrigerated environment and shipped under chain-of-custody procedures to a state-certified laboratory.

Various DPT soil samplers have been designed and manufactured by numerous companies. The main sampler types used in DPT projects include split-spoon samplers, open-tube samplers, piston samplers, and dual tube samplers.

The split-spoon sampler consists of a sample barrel that can be split into two along the length of the sampler to expose soil liners. The split-spoon sampler used without sample liners is useful for lithologic logging where soil samples will not be collected for chemical analysis.

The open-tube sampler (Figure F.2) contains soil liners and has been designed for environmental sampling within the same borehole, providing that soil sloughing or bore-wall caving is minimal. Continuous coring with the open-tube sampler begins at the ground surface with the open-ended sampler. The open-tube sampler is reinserted back down the same borehole to obtain the next core. The open-tube sampler works well in stable soils such as medium- to fine-grained cohesive materials, like silty clay soils or sediments. The open-ended samplers are commonly 3/4 to 2 in. in diameter and 2 to 5 ft in length. The simplicity of the open-tube sampler allows for rapid coring.

For discrete sampling in unstable soils, the piston sampler (Figure F.3) allows the user to drive the sampler to the selected sampling depth. The piston sampler is equipped with a piston assembly that locks into the cutting shoe and prevents soil from entering the sampler as it is driven in the existing borehole. After the sampler has reached the zone of interest, the piston is unlocked to allow the soils to push the piston as the sampler is advanced into the soil.

Dual-tube soil samplers (Figure F.4) were designed to prevent cross contamination when sampling through perched watertables. The dual-tube sampler has an outer rod that is driven ahead of the inner sampler. The outer rod provides a sealed hole from which soil samples can be recovered without the risk of cross contamination. The inner rod with sampler and outer casing are driven together to one sampling interval. The inner sampling rod is then retracted to retrieve the filled liner while the outer casing is kept in place to prevent sloughing. This procedure is then repeated to the total desired depth for sampling. Although slower than open-tube sampling, dual-tube sampling is especially useful when sampling through sloughing soils and sediments.

F.4.2.2 CPT soil sampling

Many sample collection options exist for CPT rigs, each depending on the specific medium of subsurface components. Many of the CPT soil samplers resemble their hollow-stem drill counterparts in that a split-spoon cutting tool is used for collection. Typically, a push rod with a closed tip (piston sampler) is advanced to a depth just above the desired sampling depth (Figure F.5). The cutting tip of a CPT sampler is generally opened using a spring-loaded hinge system inside the sampling string. Once the hinge is triggered, the open cutting tube is advanced to the desired depth, filling the cylindrical sample chamber with soil. The former tip is retracted during collection. This technique tends to work best in dry soil with low gravel percentages. In some over-pressured sandy environments beneath the

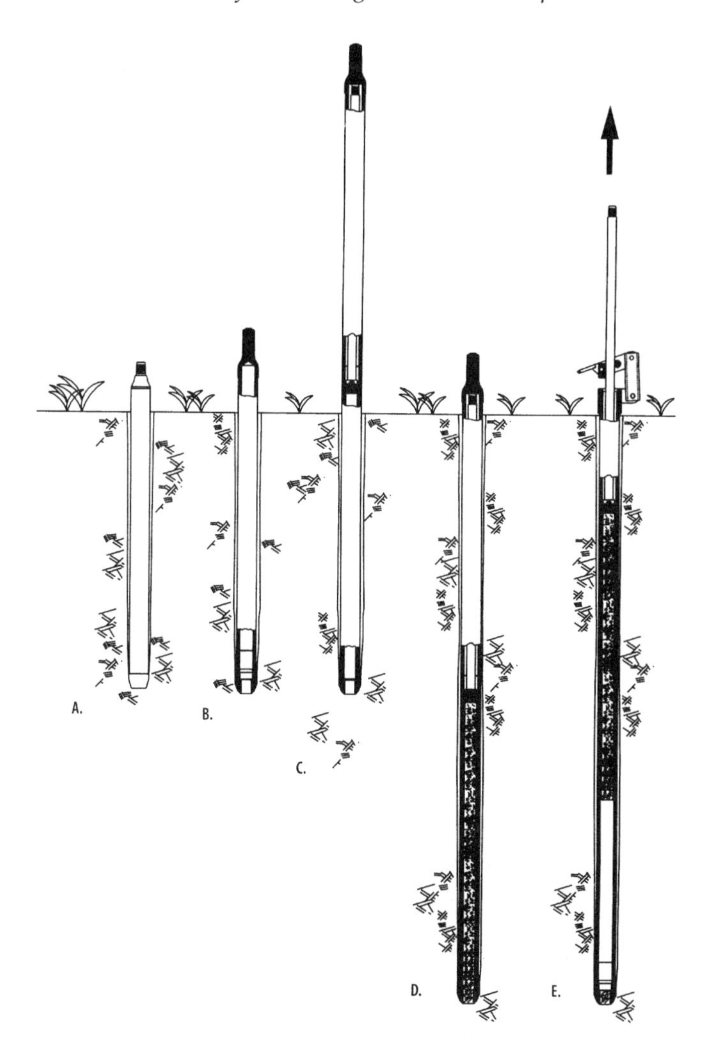

Figure F.2 Open tube soil sampler. (A.) First soil core retrieved with Macro-Core sampler. (B.) DT21 outer casing (without liner) advanced to bottom of previously cored hole. (C.) Sample liner, drive head, and inner rod placed inside casing. Outer casing section, drive bumper, and drive cap added to tool string. (D.) Tool string driven to collect soil core. (E.) Inner rod and liner (with soil core) retrieved with help of DT21 Rod Clamp Assembly. (From Geoprobe®, 1999. With permission.)

watertable, sample collection can be difficult. If relatively finer-grained material is located below the sand of interest, it is sometimes useful to space the sampling interval so that the finer material forms a plug at the bottom of the push stroke. More detailed information is described in Edelman and Holguin, 1996.

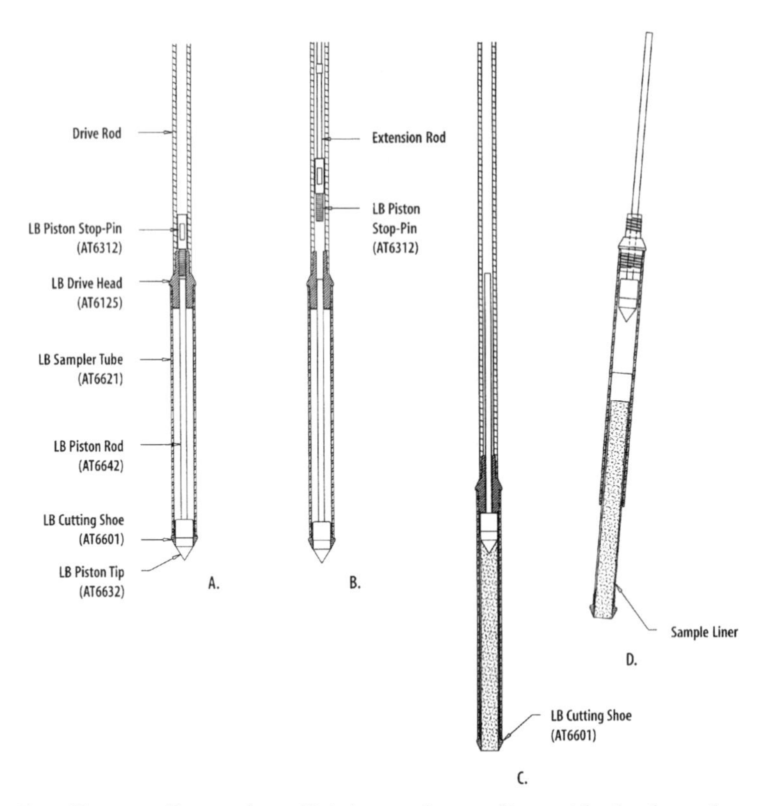

Figure F.3 Piston soil sampler. Driving and sampling with the large bore soil sampler, 1.25-in. probe rod system (using heavy-duty 9/16 in. stop-pin). (A.) Driving the Sealed Sampler. (B.) Removing the piston stop-pin in preparation for sample collection. (C.) Sample collected and ready for retraction to surface. (D.) Recovered sample in liner. (From Geoprobe®, 1999. With permission.)

F.4.2.3 DPT/CPT vapor sampling

Vapor sampling can be useful in evaluating vapor migration of the more volatile components of gasoline, such as benzene. Soil gas information can also be used to define the location and vertical and lateral extent of the residual chemicals in the vadose zone. However, due to its high solubility in water and its relatively low vapor pressure, MTBE is rarely found in the vapor phase; consequently, vapor sampling is rarely performed to identify MTBE contamination in the soil.

F.4.2.4 DPT water sampling

Some DPT probe water samplers, such as the Geoprobe® Groundwater Sampler, use a retractable or expendable drive point (Figure F.6). After driving to the zone of interest, the outer casing of the probe is raised from the borehole, exposing the underlying well screen. For a non-discrete groundwater sample, an outer casing containing open slots is used. The open-slotted tool is driven from ground surface into the water table. Groundwater is collected using an

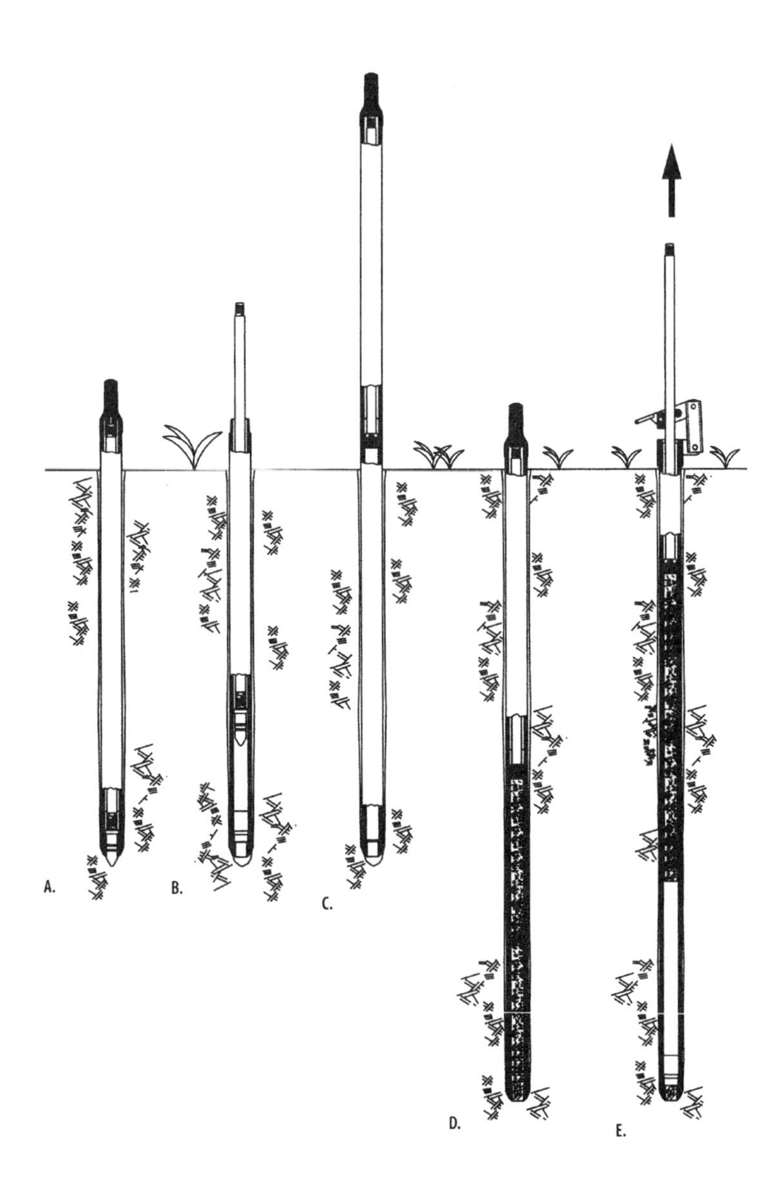

Figure F.4 Dual tube soil sampler. Driving and sampling with the dual tube soil sampler. (A.) Outer casing advanced through undisturbed soil to top of sampling interval. (B.) Solid drive tip, inner rod, and threadless drive cap seal casing as it is advanced. (C.) Solid drive tip and inner rod string removed from outer casing. (D.) Sample liner, drive head, and inner rod placed inside casing. Outer casing section, drive bumper, and drive cap added to tool string. (E.) Inner rod and liner (with soil core) retrieved. (From Geoprobe®, 1999. With permission.)

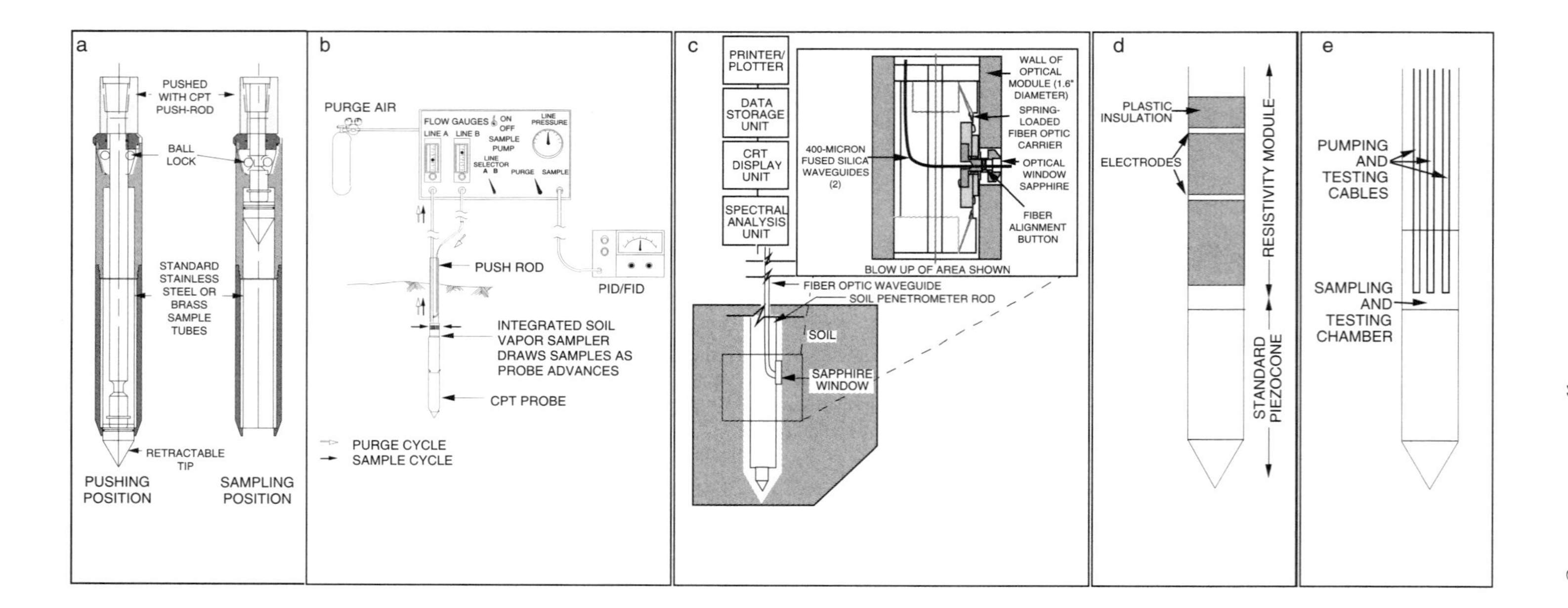

Figure F.5 CPT Soil sampling. (From Edelman and Holguin, 1996. With permission.)

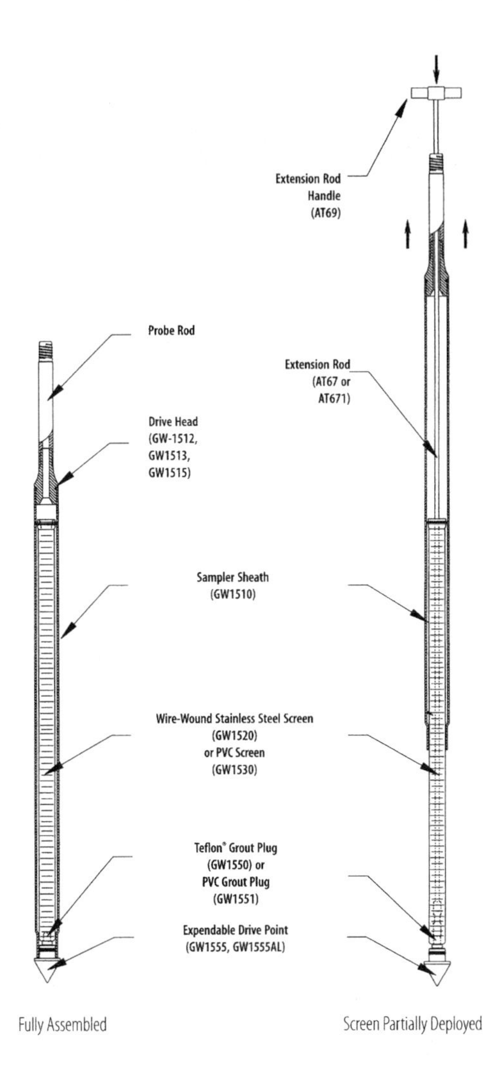

Figure F.6 Retactable Groundwater Sampler. (From Geoprobe®, 1999. With permission.)

inner tubing or smaller diameter bailer inserted into the center of the open-slotted water sampler.

F.4.2.5 CPT water sampling

Various types of sealed groundwater samplers are available for push technology, as well as the conventional rotary drilling methods.[3] These groundwater samplers include MaxiSimulProbe® (Figure F.7), HydroPunch® I and II (Figure F.8), ConeSipper®, Bat® EnviroProbe (refer to Figure F.8), and others.

Groundwater samplers typically work by either pumping or mass displacement. The ConeSipper® water sampler consists of a screened lower

Table F.2 Sampling and Monitoring Tools

Category	Sampler	Source Zone Tracking	Dissolved Plume Tracking	Geologic Characterization	Installation Rig
One-time sampling tools					
Sealed Screen Sampler	BAT	4	4	3	HSA, CPT, MR, DWP, ARCH
Sealed Screen Sampler	HPI	3	2	2	HSA, CPT, MR, DWP, ARCH
Sealed Screen Sampler	HPII	3	2	2	HSA, CPT, MR, DWP, ARCH
Sealed Screen Sampler	H2VAPE	4	4	3	HSA, CPT, MR, DWP, ARCH
Sealed Screen Sampler	MXSP	4	4	4	HSA, CPT, MR, DWP, ARCH
Sealed Screen Sampler	MNSP	4	4	3	
Vertical Profilers	WGP	4	3	2	Probe, CPT
Vertical Profilers	GGP	4	3	2	Probe, CPT
Vertical Profilers	CS	4	4	4	CPT
Long-term monitoring tools					
Well points (0.5 to 1" dia. Piez.)	Bailer, pump	3	3	1	Probe, HSA, CPT
Multi-Level Wellpoints	Bailer, pump	3	3	2	Probe, HSA, CPT
Monitoring Wells	Bailer, pump	3	3	1	HAS
Multi-Level Wells	Bailer, pump	3	3	1	HAS

NOTES:
4 = best application
3 = good application
2 = fair application
1 = poor appilcation
NA = not applicable

RIGS:
HSA = Hollow Stem Auger Rig
CPT = Cone Pentrometer Testing Rig
Probe = Direct Push Technology Rig
MR = Mud rotary
DWP = Dual Wall Percussion
ARCH = Air Rotary Casing Hammer

SAMPLERS:
BAT = Bat EnviroProbe
HPI = HydroPunch I
HPII = HydroPunch II
WGP = Waterloo Groundwater Profiler
GGP = Geoprobe Groundwater Profiler
CS = Cone Sipper
H2VAPE = H2VAPE Simulprobe
MXSP = Maxi Probe Simulprobe
MNSP = Mini Probe Simulprobe

(Table prepared by N. Heller and J. Jacobs.)

chamber and an upper collection chamber. Two small-diameter Teflon® tubes connect the upper collection chamber to a control panel in the truck. The truck ballast and hydraulic rams are used to push the sampler to a predetermined depth. While the probe is being advanced, nitrogen, under relatively high pressure, is supplied to the collection chamber via the pressure/vacuum Teflon® tube in order to purge the collection chamber and prevent groundwater from entering the chamber before the probe reaches the desired depth. Once the desired depth is reached, the pressurized nitrogen is shut off, excess nitrogen pressure is removed from the 100 milliliter upper collection chamber, and the chamber fills with groundwater. Finally, nitrogen, under relatively low pressure, is supplied to the collection chamber to gently displace the water and slowly push the water to the surface through the small-diameter Teflon® sampling tube. Slow sampling and low pressures are critical for sampling water containing volatile constituents. Sample collection times range from 20 min. to 2 h, depending on the soil type adjacent to the sampling port. The groundwater samples are placed directly into 40 milliliter volatile organic analysis (VOA) containers. The VOA vials are labeled, logged, and placed into a chilled cooler pending delivery to an analytical laboratory.

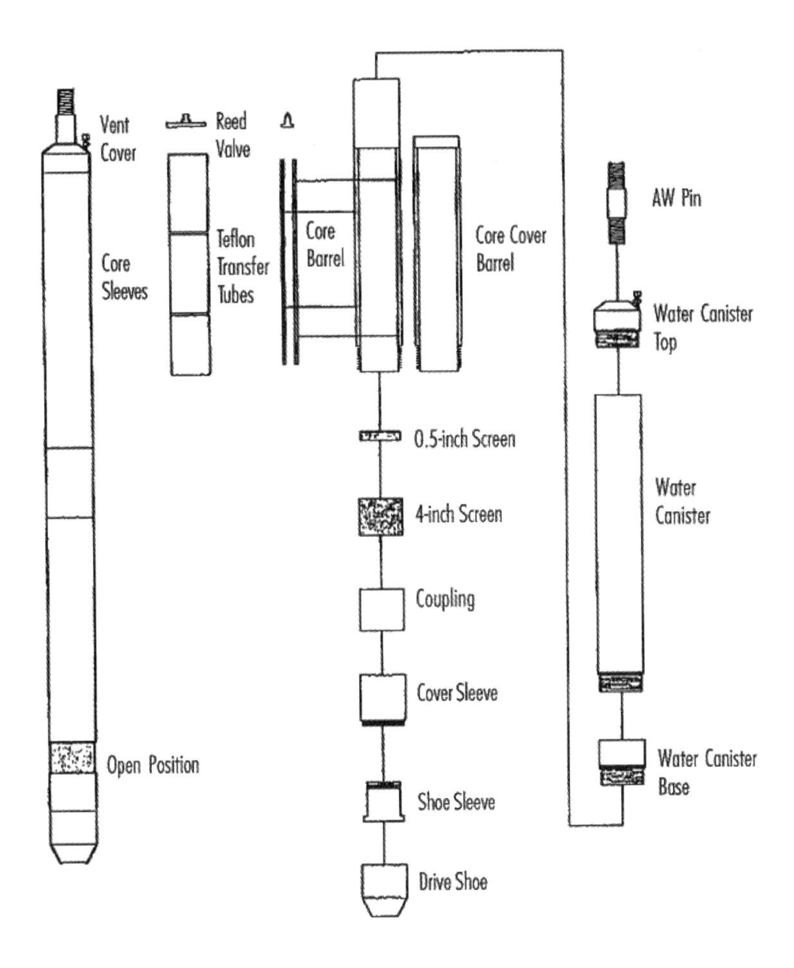

Figure F.7 MaxiSimulProbe. (From Heller and Jacobs 2000. With permission.)

F.5 Conventional drilling methods

The following section presents some of the more conventional drilling techniques.

F.5.1 Hollow-stem auger drilling

Hollow-stem continuous flight auger drilling techniques are commonly used for subsurface environmental projects. The augers and the hollow-stem auger rig are shown in Figures F.9 and F.10, respectively. Hollow-stem augers consist of a series of continuous, interconnected hollow auger flights, usually 5 to 10 ft in length. Typical hollow-stem augers for use in the environmental field have inner diameters (ID) of 6 to 10.25 in., which create 2- and 4-in.

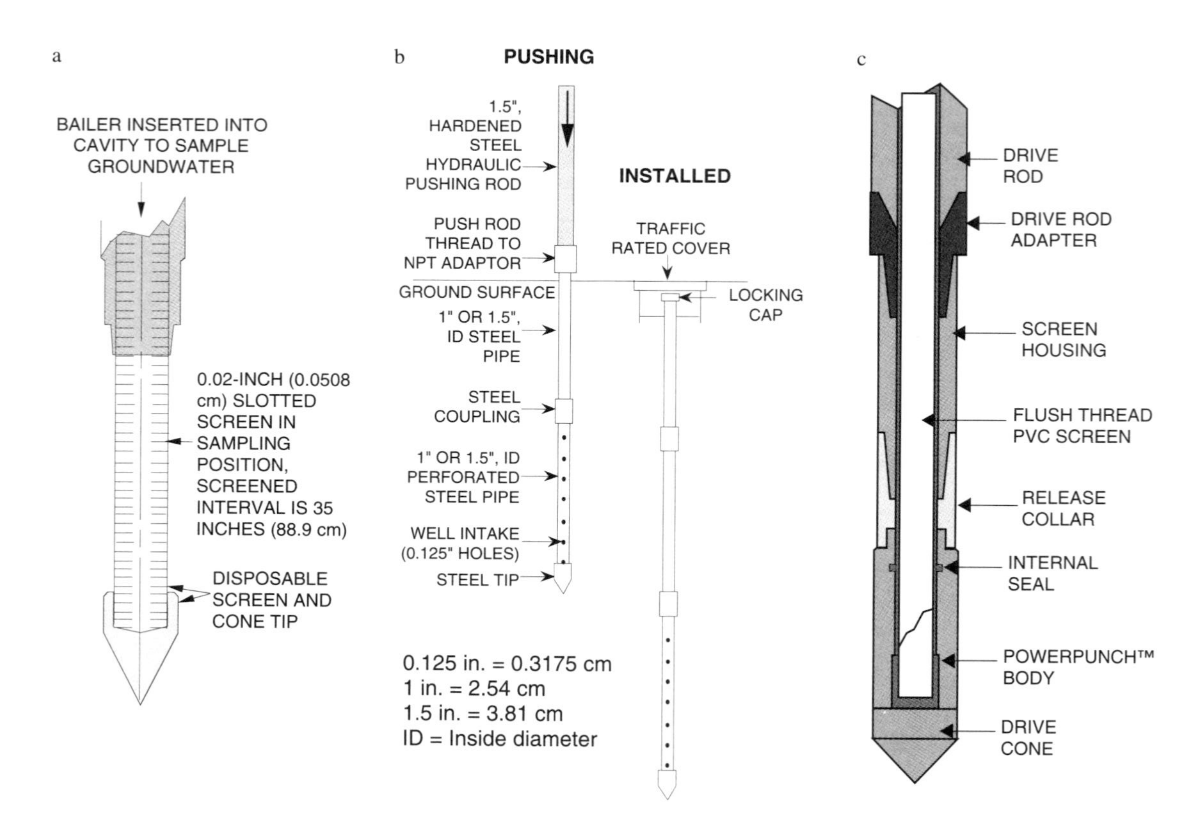

Figure F.8 Various CPT groundwater samplers. (From Edelman and Holgun, 1996. With permission.)

diameter wells, respectively. The hollow-stem flight augers are hydraulically pressed downward and rotated to initiate drilling. Soil cuttings are rotated up the outside of the continuous flighting in the borehole annulus. A center rod with plug and pilot bit are mounted at the bottom. The plug is designed to keep soil from entering the mouth of the lead auger while drilling.

Upon reaching the desired sampling depth, the center rod string with plug and pilot bit attached is removed from the mouth of the auger and replaced by a soil sampler.

F.5.2 Cable-tool drilling

Cable-tool drilling is the oldest drilling technique available and is not used widely or frequently in the environmental field (Figure F.11). The exception is the use of cable-tool drilling in glacial environments that contain large cobbles in the Pacific Northwest portion of the U.S. Cable-tool rigs, called percussion or spudder rigs, operate by repeatedly lifting and dropping the heavy string of drilling tools in the borehole, crushing larger cobbles and rocks into smaller fragments. Water is needed to create a slurry at the bottom of the borehole. When formation water is not present, water is added to form a slurry. The addition of large volumes of water into the formation to create the slurry may degrade and compromise the quality of environmental samples — it is partly for this reason that this technique is not used widely or frequently in the environmental field.

With use of the cable tool, the hole is continuously cased with an unperforated, 8-in.-diameter steel casing with a drive shoe. The casing is attached on top by means of a rope socket to a cable that is suspended through a pulley from the mast of the drill rig. The process of driving the casing downward into the subsurface about 3 to 5 ft, followed by bailing, is referred to as "drive and bail". Cuttings are removed by periodically bailing the borehole; thus, water must be added to the borehole to create a slurry when drilling in nonwater-bearing material. When drilling under saturated conditions, the drive and bail process allows for the collection of depth-discrete groundwater samples, that can be field screened using a portable organic vapor analyzer in the gas chromatograph mode, to assess the potential presence of dissolved petroleum hydrocarbon constituents in groundwater (Testa, 1994).

F.5.3 Rotary drilling

Rotary drilling techniques include direct mud rotary, air rotary, air rotary with a casing driver, and dual-wall reverse circulation. With direct mud rotary, drilling fluid is pumped down through the bit at the end of the drill rods, then is circulated up the annular space back to the surface (Figure F.12). The fluid at the surface is routed via a pipe or ditch to a sedimentation tank or pit, then to a suction pit where the fluid is recirculated back through the drill rods. Air rotary drilling is similar to direct mud rotary drilling, except that air is used as a circulation medium instead of water. Although the air helps cool the bit, small quantities of water or foaming surfactants are used to facilitate sampling. In unconsolidated deposits, direct mud or air rotary can be used, providing that a casing is driven as the drill bit is advanced. In dual-wall reverse circulation, the circulating medium (mud or air) is pumped downward between the outer casing and inner drill pipe, out through the drill bit, then up the inside of the drill pipe.

Rotary drilling techniques are commonly limited to consolidated deposits of rocks and typically not used in subsurface environmental studies due to their limited sampling capabilities. Sample integrity is questionable since the added water, mud, or surfactant may chemically react with the formation water. In addition, thin water-bearing zones are often missed with the use of these techniques. With mud rotary, the mud filter cake that develops along the borehole wall may adversely affect permeability of the adjacent formation materials. With air rotary, dispersion of potentially hazardous and toxic particulates in the air during drilling is a concern. Air rotary drilling techniques are fast and, where the subsurface geology is relatively well-characterized or a resistant stratum (i.e., overlying basalt flows or conglomerate strata) exists at shallow levels within the vadose zone and above the depth of concerns, utilization of air rotary to a limited depth followed by another more suitable drilling technique may be worth considering (Testa, 1994).

F.6 Conventional soil sampling

Soil sampling is achieved by passing a smaller diameter drill rod into the hollow-stem auger with a soil sampler attached at the bottom. The sampler is typically either a thin-wall or modified split-spoon sampler with stainless steel or brass sample tubes. Samples can be continuously retrieved, although in the environmental field, soil samples are typically collected at 5-ft intervals, or at significant changes in lithology. In addition, soil samples are usually collected at intervals of obvious contamination in order to develop a complete profile of soil contamination.

The sampler is driven into the soil at the desired sampling interval ahead of the auger bit. A standard California modified split-spoon sampler is lined with three stainless steel or brass tubes each 6 in. in length. Typically, the upper 6-in. sample tube is used to collect a sample for lithologic description and physical testing (i.e., permeability, sieve analysis, etc.). The middle sample tube is used to collect a sample for field screening for hydrocarbon or solvent vapors or other constituents of concern, and the bottom sample tube is used to collect a sample to send to a chemical laboratory for analysis.

Using a 140-lb hammer with a 30-in. drop, a standard penetration test is performed. The number of blows required to drive the sampler for each 6-in. interval is recorded on a boring log. The blow count is used to determine consistency and soil density information. After the sampler is driven 18 in., the sampler is extracted, and the liner containing the sample is removed.

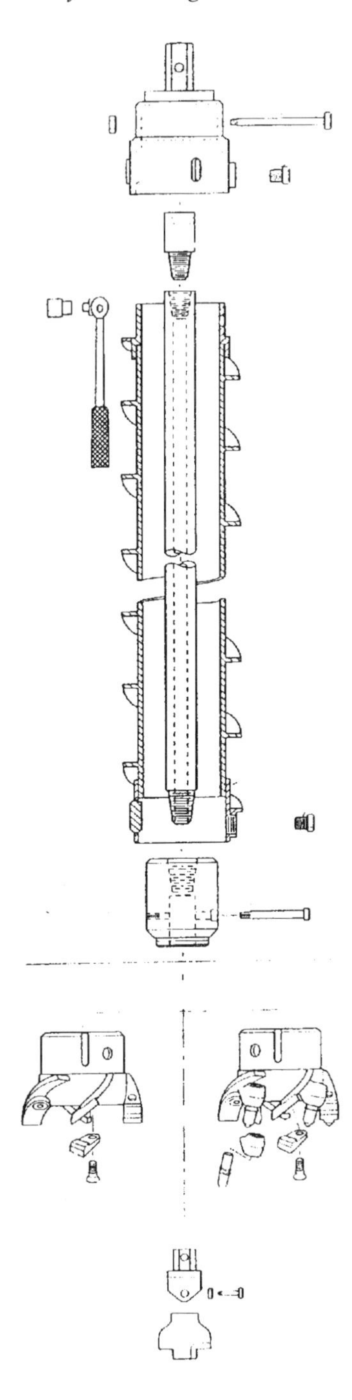

Figure F.9 Hollow stem auger with cutter heads. The center rod and pilot bit operate in the center of the auger. (From Foremost Mobile Drill and Johnson Well Screen. With permission.)

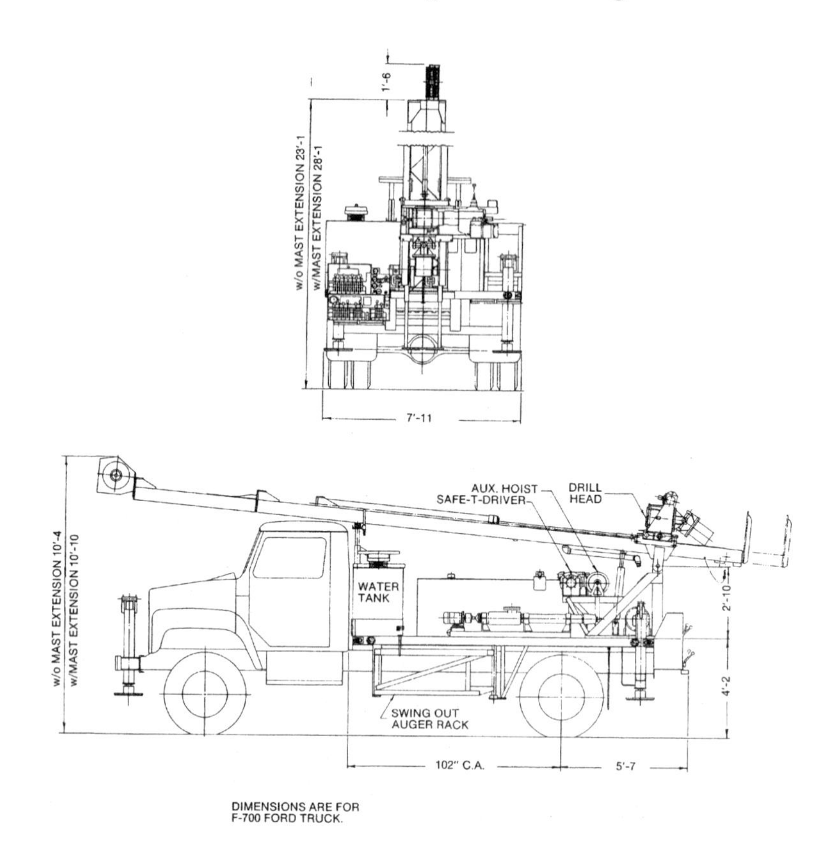

Figure F.10 B-53 Hollow stem auger rig. (From Foremost-Mobile Drill. With permission.)

Boreholes are either filled in by grouting using a tremie pipe or converted into a groundwater monitoring well. Soil cuttings are generally not used to backfill borings in environmental projects due to the potential for cross contamination.

F.7 Installation of groundwater monitoring wells and well points

F.7.1 Well installation by rotary drilling rig

The purpose of a monitoring well network is to define groundwater quality and movement and to accomplish specific study objectives. The borehole for a monitoring or extraction well is frequently drilled using a truck-mounted, continuous flight, hollow-stem auger drill rig. The borehole diameter for

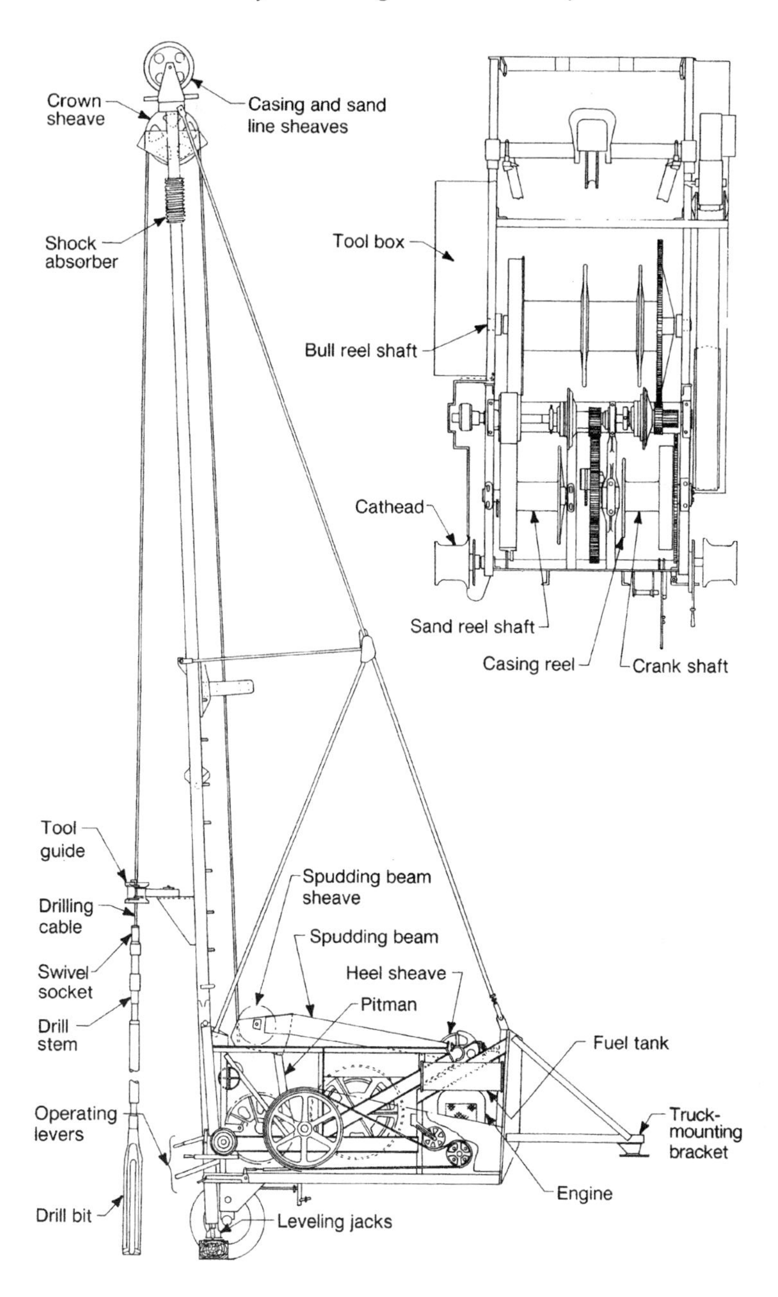

Figure F.11 Cable-tool drilling: the percussion action is transferred to the drill line by the vertical motion of the spudding beam. (From Driscoll, F., *Groundwater and Wells*, Johnson Well Screen, 1994. With permission.)

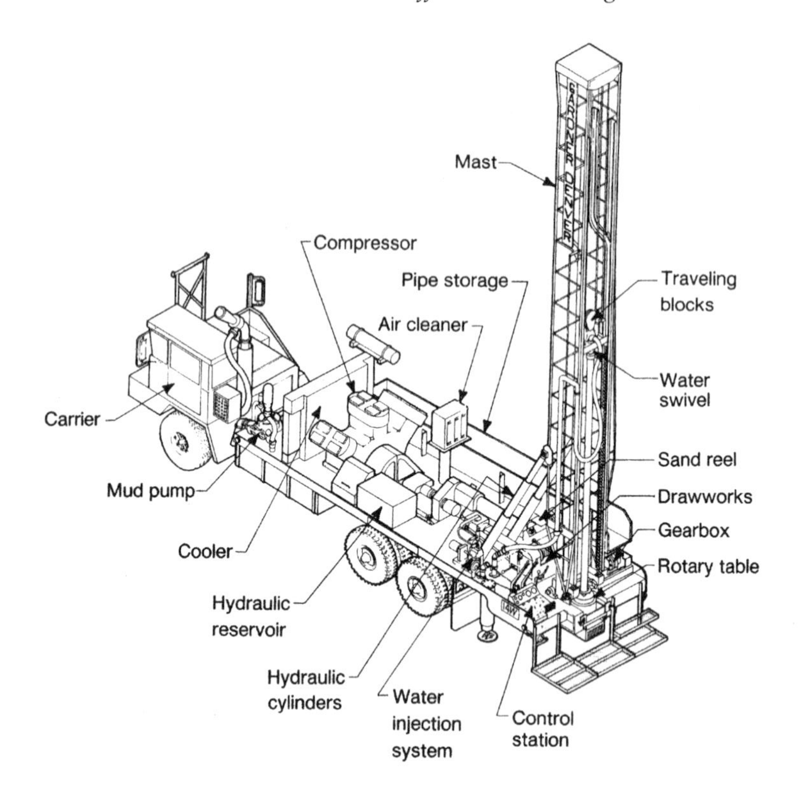

Figure F.12 Schematic diagram of a direct rotary rig. (From Driscoll, F., Groundwater and Wells, Johnson Well Screen, 1994. With permission.)

monitoring is a minimum of 2 to 4 in. larger than the outside diameter of the well casing, in accordance with appropriate regulatory guidelines. The hollow-stem auger allows for minimal interruption of drilling, while permitting soil sampling at the desired intervals.

For hollow-stem auger drilling, the auger will remain in the borehole while the well casing is set, to prevent caving. The well is cased with blank and factory-slotted, threaded schedule-40 polyvinyl chloride (PVC) pipe (Figure F.13). The slotted casing is placed from the bottom of the borehole to the top of the aquifer. The blank casing will extend from the slotted casing to approximately ground surface. The slots are generally 0.010 in. or 0.020 in. wide by 1.5 in. long, with approximately 42 slots per linear foot. Slot sizes are determined by previous well installations in the area, or by a grain size sieve analysis. A threaded PVC cap is fastened to the bottom of the casing. Centering devices may be fastened to the casing to assure even distribution of filter material and grout within the borehole annulus. The well casing is typically steam cleaned prior to installation to insure that no machine oils or other hazards exist on the casing surfaces.

After setting the casing within the auger, sand or gravel filter material is poured into the annulus to fill from the bottom of the boring to 1 ft above the slotted interval. The auger is slowly removed in advance of the filter material. After the filter material is in place, the auger is rotated out of the borehole and completely removed. One to two ft of bentonite pellets is placed above the filter material, and then hydrated with deionized water. The bentonite pellets are placed to prevent the grout and surface contaminants from reaching the filter material. Neat cement, a common grout for sealing environmental wells, contains approximately 5% bentonite powder. The grout is tremied into the annular space from the top of the bentonite plug to the surface. The grout need not be tremied if the top of the bentonite plug is less than 5 ft below ground surface, and the interval is dry. Approved grout mixtures and grouting techniques may vary depending on local conditions and regulations.

A lockable PVC cap is typically placed on the wellhead. A traffic-rated flush-mounted steel cover is installed around a wellhead located in traffic areas. A steel stove pipe is usually set over a wellhead in landscaped areas. The flush-mounted cover box or stovepipe monument contacts the grout. Grout fills the space between the monument and the sides of the borehole. The monument and grout surface seal is set at or above grade so that drainage is away from the monument.

F.8. Groundwater sampling protocols

MTBE water sampling from monitoring wells

For newly installed wells, a minimum of 24 hr is allowed to lapse between well development and sampling. If well development methods are used which may introduce air into the groundwater (i.e., air lifting), a minimum of 72 hr should lapse prior to sampling. The waiting period ensures that any air that may have been introduced by development techniques dissipates. If free product (gasoline floating on the surface of the groundwater in the well) is detected during well sampling, analysis for dissolved product of groundwater at the interface is not performed. A product sample is collected for source identification. If all free product cannot be removed, an internal-specific sampling device may be utilized to collect a groundwater sample from below the layer of free product.

Cross-contamination between wells is avoided by careful decontamination procedures. Purging proceeds from the least to the most contaminated well, if this is known or indicated by field evidence. The well is purged until indicator parameters (pH, temperature, and conductivity) are stabilized (within 10%). A minimum of three wetted casing volumes is removed by bailing or pumping. If the groundwater aquifer is slow to recharge, purging

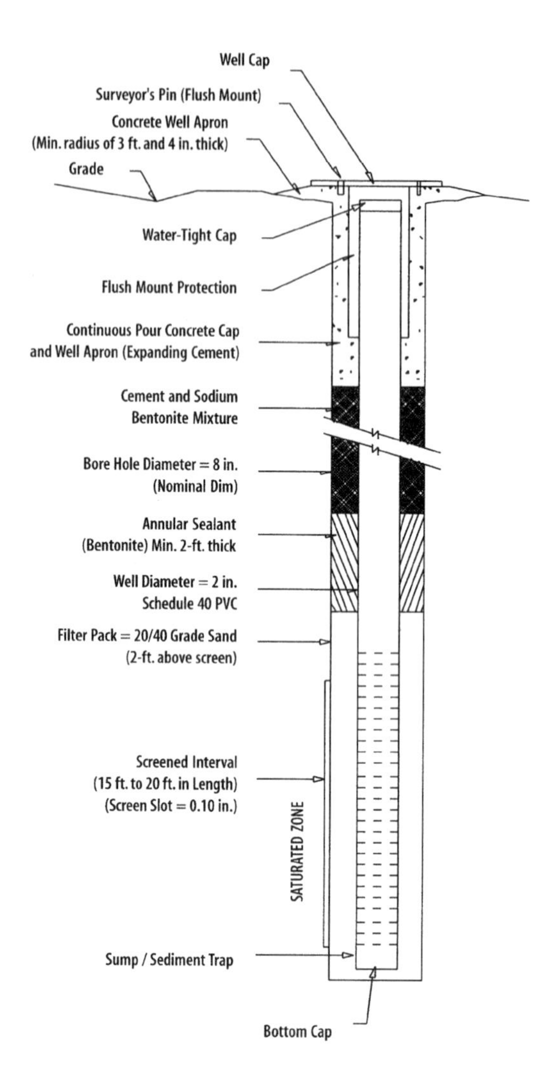

Figure F.13 Monitoring well construction diagram. (From Jacobs, 1999. With permission.)

will continue until the well is pumped dry. Once the well is sufficiently purged, a sample may be collected after the water level in the well has achieved 80% of its initial volume. Where water level recovery is slow, the sample may be collected after two hr. Samples are collected using a disposable polyethylene bailer with a bottom siphon and nylon cord. If a Teflon or stainless steel bailer is used, it is decontaminated between wells. Samples are transferred to clean laboratory-supplied containers. Cross-contamination in groundwater samples may provide a false positive, while poor handling may allow for volatilization of the BTEX components or a decrease from the actual in situ MTBE concentrations.

Endnotes and references

Portions of this text were adapted from contributions by the author to *The Standard Handbook of Environmental Science, Health and Technology*, a McGraw-Hill book. In addition, portions of this text were adapted from contributions by Stephen M. Testa to *Geological Aspects of Hazardous Waste Management*, a Lewis Publishers/CRC Press book. Geoprobe® is a Registered Trademark of Kejr Engineering, Inc.; Teflon® is a Registered Trademark of E.I. du Pont de Nemours & Company; Maxi SimulProbe®, Mini SimulProbe®, and H2-Vape® are Registered Trademarks of BESST, Inc.; ConeSipper® is a Registered Trademark of Vertek Division of Applied Research Associates. HydroPunch® I and II are Registered Trademarks of QED Environmental; Waterloo Profiler® is a Registered Trademark of Solinst; Bat®EnviroProbe is a Registered Trademark of Hogentogler.

American Society for Testing Materials (ASTM), Standard Practice for Description and Identification of Soils (Visual-Manual Procedure), Method D 2488-84, December 1984.

Blegen, R.P., Hess, J.W. Denne, J.E., 1988, Field Comparison of Groundwater Sampling Devices, National Water Well Association Outdoor Action Conference in Las Vegas, NV, May, 1988.

California Department of Water Resources (DWR), California Well Standards, Bulletin 74-81, 1981.

DWR, 1990; California Well Standards, Bulletin 74-90, January 1990.

California Regional Water Quality Control Board (RWQCB), Leaking Underground Fuel Tank (LUFT) Field Manual: Guidelines for Site Assessment, Cleanup, and Underground Storage Tank Closure, October 1989.

RWQCB, Tri-Regional Board Staff Recommendations for Preliminary Investigation and Evaluation of Underground Tank Sites, August 10, 1990.

Driscoll, F. G., *Groundwater and Wells*, 2nd Ed., Johnson Filtration Systems, St. Paul, pp. 268–339, 1986.

Edelman, S.H., and Holguin, A.R., 1996, *Cone Penetrometer Testing for Characterization and Sampling of Soil and Groundwater, Sampling Environmental Media*, ASTM STP 1282, James Howard Morgan, Ed., American Society for Testing and Materials, Philadelphia, PA, 1996.

Geoprobe Systems, 98-99 Tools and Equipment Catalog, Kejr Engineering, Inc. Salina, KS, 1997.

Heller, N., and Jacobs, J.A., Everything You Wanted to Know About Depth Discrete Groundwater Samplers for Vertical VOC Profiling, 2000 (in press).

Jacobs, J.A. and Loo, W.W., Direct Push Technology Methods for Site Evaluation and In Situ Remediations, Hydrovisions, July/August, Groundwater Resources Association of California, 1994.

Jacobs, J.A., Monitoring well construction and sampling techniques, in *The Standard Handbook of Environmental Science, Health and Technology*, Jay Lehr, Ed., McGraw Hill Publishing Company, 2000 (in press).

Jacobs, J.A., (in press), Direct Push Technology Sampling Methods, in *The Standard Handbook of Environmental Science*, Health and Technology, ed. Jay Lehr, McGraw Hill Publishing Company.

Nichols, E., Einerson, M., and Beadle, S., Strategies for Characterizing Subsurface Releases of Gasoline Containing MTBE, API Publication No. 4699, February, American Petroleum Institute, Washington, D.C., 2000.

Sisk, Steven J., *NEIC Manual for Groundwater/Subsurface Investigations at Hazardous Waste Sites*, EPA-330/9-81-002, U.S. Environmental Protection Agency, Denver, Colorado, July, 1981.

Testa, S. M., *Geologic Aspects of Hazardous Waste Management*, CRC Press, Boca Raton, Florida, 145–187, 1994.

Weaver, J.W., Charbeneau, R.J., Tauxe, J.D., Lien, B.K., and Provost, J. B., The Hydrocarbon Spill Screening Model (HSSM) Volume 1: User's Guide, EPA/600/R-94/039a, Robert S. Kerr Environmental Research Laboratory, Office of Research and Development, U.S. Environmental Protection Agency, Ada, Oklahoma, 1994.

Weaver, J.W., Haas, J.E., and Sosik, C. Characteristics of gasoline releases in the water table aquifer of Long Island, in *Proc. NGWA/API Conf. Petrol. Hydrocarb. Organ. Chem. Groundwater: Prevention, Detect., Remed.*, Houston, Texas, Nov. 17–19, 260–261, 1999.

Zemo, D.A., T.A. Delfino, J.D. Gallinatti, V.R. Baker, and L.R. Hilpert, Field comparison of analytical results from discrete-depth groundwater samplers, *Groundwater Monitoring Rev.*, Winter, 133–141, 1995.

[1] For more information on investigating MTBE spills and characterization of aquifers, please refer to Weaver et al., 1994, 1999, and Nichols et al., 2000.

[2] Sisk, 1981.

[3] Blegen et al., 1988; Zemo et al., 1995; Heller and Jacobs, personal communication.

APPENDIX G

Synthesis, properties, and environmental fate of MTBE and oxygenate chemicals

G.1 Formation of MTBE: chemical basis for physical properties

Methyl *tertiary*-butyl ether (MTBE), as discussed in several chapters of the main body of this book, is a biorefractory compound that is not readily adsorbed to organic material, is oxidized neither biologically nor chemically, and is not readily removed from water through air stripping. The chemical structure of the compound, determined during its synthesis, is the controlling factor that determines all of these properties.

The synthesis of MTBE, as commercially implemented, is achieved through the addition of methanol to isobutylene, an olefin (or alkene). The reaction occurs in the presence of an acidic catalyst, which is necessary to form the reactive intermediate compound, the carbo-cation, as indicated in Figure G.1.

The reaction is formally known as the electrophilic addition of methanol to isobutylene. As can be observed in Figure G.1, there are two possible orientations of the molecule that can result from the reaction. The chemically preferred orientation is the one shown, the *tertiary*-butyl methyl ether. The tertiary structure is overwhelmingly preferred in the reaction pathway due to the stability of the tertiary carbo-cation.

The preference of hydrocarbons to orient in the tertiary structure as a result of the formation of the most stable carbo-cation is known as Markovnikov's rule. In other words, it is the tendency of the inner structure of the hydrocarbons to conform to the most stable orientation that produces the tertiary carbon structure. That structure, in turn, results in the resistance of the molecule to biodegradation and chemical oxidation.

The presence of the oxygen atom in the molecule, as an ether linkage, provides a site for hydrogen bonding between water and MTBE. The energy

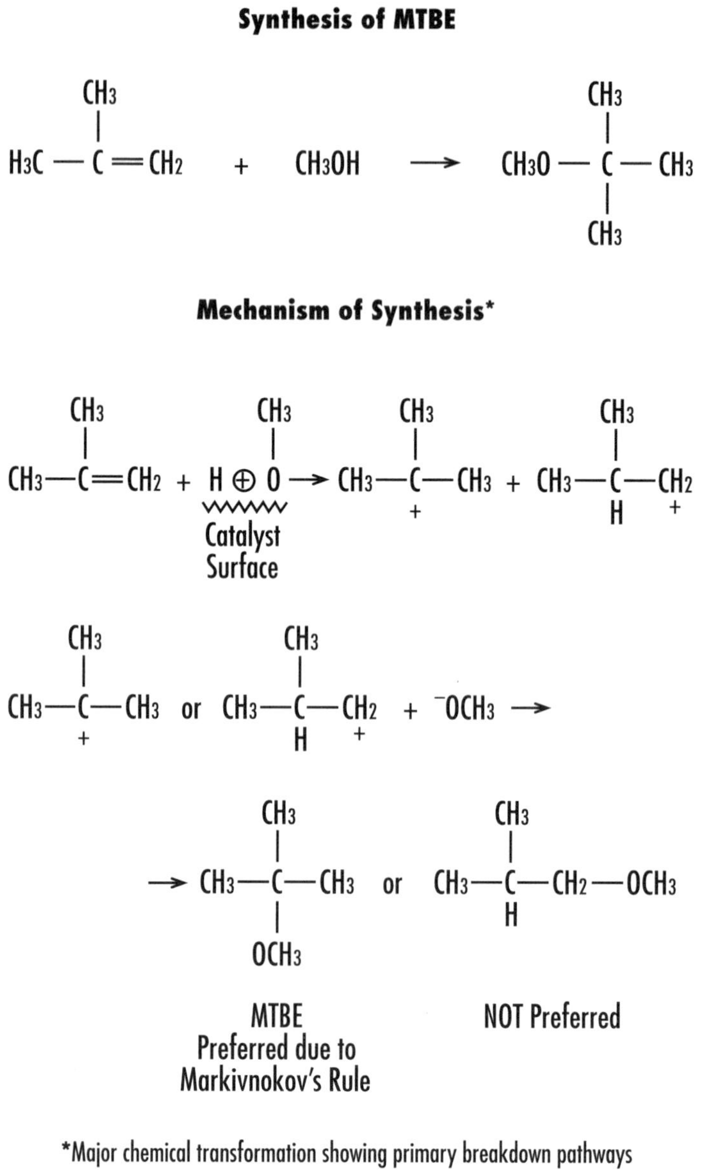

Synthesis of MTBE

$$H_3C-\overset{\displaystyle CH_3}{\overset{|}{C}}=CH_2 \quad + \quad CH_3OH \quad \longrightarrow \quad CH_3O-\overset{\displaystyle CH_3}{\overset{|}{\underset{\displaystyle CH_3}{\underset{|}{C}}}}-CH_3$$

Mechanism of Synthesis*

$$CH_3-\overset{\displaystyle CH_3}{\overset{|}{C}}=CH_2 + \underset{\text{Catalyst Surface}}{H \oplus \overset{\displaystyle CH_3}{\overset{|}{O}}} \rightarrow CH_3-\overset{\displaystyle CH_3}{\overset{|}{\underset{+}{C}}}-CH_3 + CH_3-\overset{\displaystyle CH_3}{\overset{|}{\underset{H}{C}}}-\underset{+}{CH_2}$$

$$CH_3-\overset{\displaystyle CH_3}{\overset{|}{\underset{+}{C}}}-CH_3 \text{ or } CH_3-\overset{\displaystyle CH_3}{\overset{|}{\underset{H}{C}}}-\underset{+}{CH_2} + {}^{-}OCH_3 \rightarrow$$

$$\rightarrow CH_3-\overset{\displaystyle CH_3}{\overset{|}{\underset{\displaystyle OCH_3}{\underset{|}{C}}}}-CH_3 \text{ or } CH_3-\overset{\displaystyle CH_3}{\overset{|}{\underset{H}{C}}}-CH_2-OCH_3$$

MTBE
Preferred due to
Markivnokov's Rule

NOT Preferred

*Major chemical transformation showing primary breakdown pathways

Figure G.1 Synthesis of MTBE.

of this bond is not high but is higher than that observed for other molecules of environmental significance, such as the chlorinated solvents. The amount of energy required to break this bond helps explain why the compound is difficult to air strip and separate from aqueous solution.

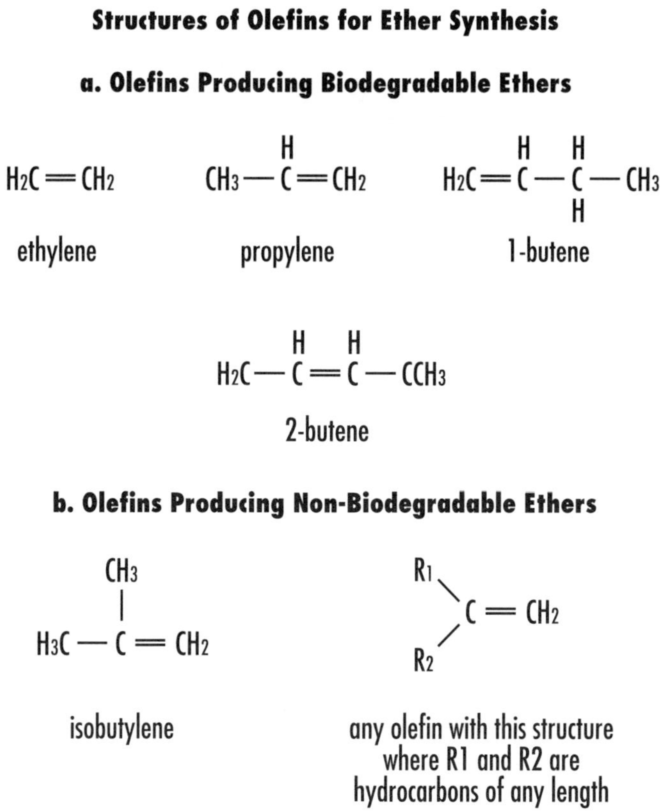

Structures of Olefins for Ether Synthesis

a. Olefins Producing Biodegradable Ethers

$H_2C{=}CH_2$ ethylene

$CH_3{-}CH{=}CH_2$ propylene

$H_2C{=}CH{-}CH_2{-}CH_3$ 1-butene

$H_2C{-}CH{=}CH{-}CCH_3$ 2-butene

b. Olefins Producing Non-Biodegradable Ethers

$H_3C{-}C(CH_3){=}CH_2$ isobutylene

$(R_1)(R_2)C{=}CH_2$ any olefin with this structure where R1 and R2 are hydrocarbons of any length

Figure G.2 Structures of olefins.

G.2 Potential oxygenate alternatives to MTBE

There are many oxygenated hydrocarbons that are potential alternatives to MTBE. Methanol is the alcohol used in the synthesis of MTBE; other alcohols, however, are also commercially available. Ethanol can be efficiently synthesized from ethylene, as can the two propyl alcohols and the four butyl alcohols. There are eight possible alternative compounds that can go into forming the alcohol/ether part of the molecule. The structures of the olefins that go into synthesis of ethers are illustrated in Figure G.2.

The alternative compounds that could potentially substitute for isobutylene and go into forming the hydrocarbon portion of the molecule are also quite varied. In fact, there is no limiting factor for why a single hydrocarbon (an alkene such as isobutylene) must be used in the synthesis.

Olefins are produced in refineries through the catalytic cracking of naphthas. While the conditions in the catalytic cracking process are suitable for isomerization of the alkenes produced, some of the compounds produced

Table G.1 Properties of Selected Oxygenates and Aromatics

Compound	CAS No.	Vapor Pressure mmHg@25°C	Molecular Weight	Henry's Constant Atm-m³/mol	Octane No. (MON)	Boiling Point °C	Log K_{ow}	K_{oc}	Specific Gravity
MTBE	1634-04-4	2.45	88.15	4.40E-04	117.00	55.2	1.2	16.22	0.744
TAME	994-05-08	88.3	102.18	1.27E-03	112.00	86		18.62	0.77
Benzene	71-43-2	0.76	78.11	5.43E-03	114.80	80.1	2.12	83	0.88
Toluene	108-83-3	28.4	92.13	5.94E-03	103.50	110.6	2.73	300	0.887
p-xylene	106-12-3	8.7	106.17	7.68E-03		137	3.15		0.861
ethanol	64-17-5	0.49	46.07	5.13E-06	102.00	78	-3.2	2.2	0.79
methanol	67-66-1	121.6	32.04	4.42E-04	105.00	64.7	-0.77	1.211	0.79
IPA	67-63-0	45.8	60.09		98.00	82			0.79
TBA	75-65-0	0.42	74.12	1.19E-05	100.00	82.9	0.35	37.15	0.79

can also be suitable for use in ether manufacture, i.e., those that do not contain the tertiary structure that can cause environmental problems. Examples of the candidate olefins and alcohols and the potential end products are provided in Figure G.2.

G.3 *Properties of the selected oxygenates and aromatics*

As indicated above, the physical properties of the ether oxygenates are determined by their molecular structure. Table G.1 provides a listing of the pertinent properties of selected oxygenates and aromatics.

G.4 *Degradation of MTBE*

Although MTBE is generally described as a biologically recalcitrant molecule, there are conditions under which the molecule can oxidize. The major factors in establishing those conditions are the availability of a population of bacteria acclimated to MTBE as a food source and the presence of an oxidant, such as hydrogen peroxide.

A proposed mechanism for the oxidative destruction of MTBE is presented in Figure G.3. *Tertiary*-butyl alcohol (TBA) is the major intermediate compound formed from the cleavage of the ether bonds in MTBE. Further reaction with the oxidant results in an effect on the primary hydrogens of TBA, and also results in unstable intermediate compounds, which are easily and further oxidized to carbon dioxide and water.

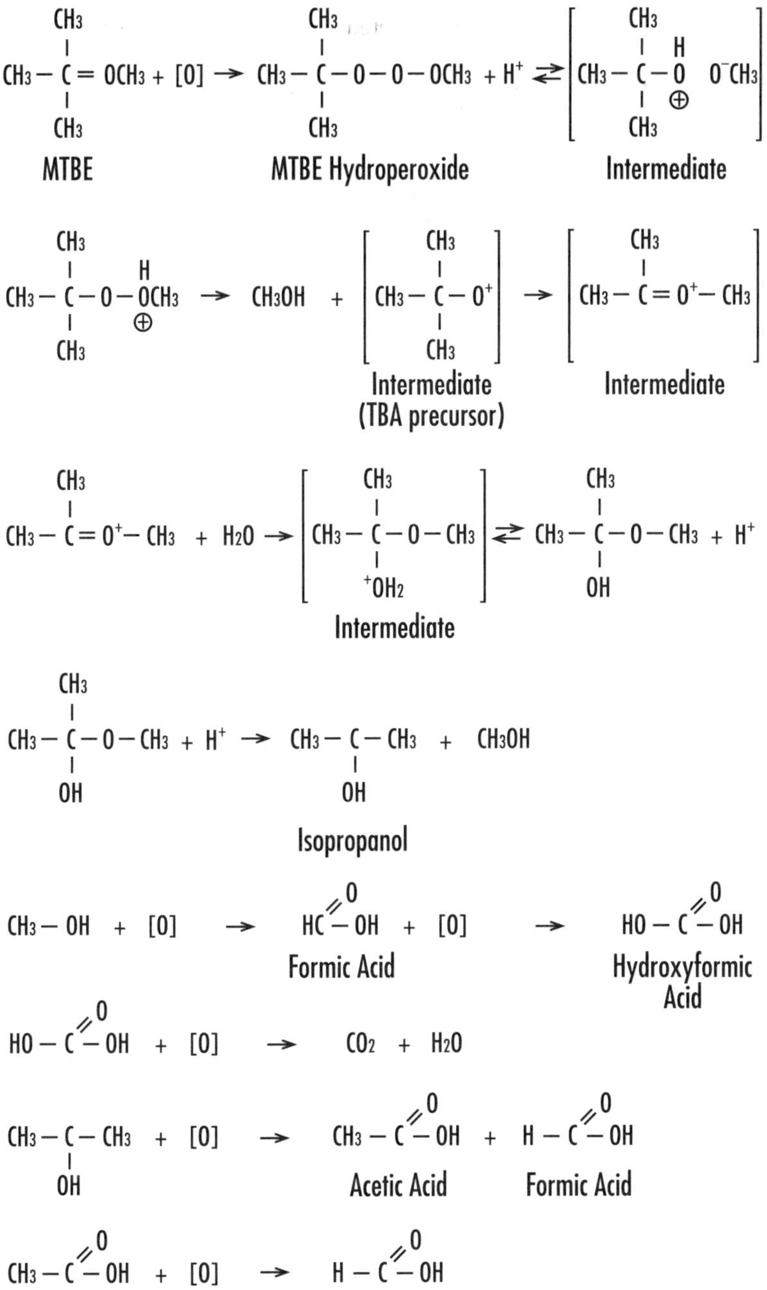

$$CH_3-C(CH_3)(CH_3)=OCH_3 + [O] \rightarrow CH_3-C(CH_3)(CH_3)-O-O-OCH_3 + H^+ \rightleftarrows CH_3-C(CH_3)(CH_3)-O^{\oplus}(H)\;\;O^-CH_3$$

MTBE — MTBE Hydroperoxide — Intermediate

$$CH_3-C(CH_3)(CH_3)-O-O^{\oplus}(H)CH_3 \rightarrow CH_3OH + CH_3-C(CH_3)(CH_3)-O^+ \rightarrow CH_3-C(CH_3)=O^+-CH_3$$

Intermediate (TBA precursor) — Intermediate

$$CH_3-C(CH_3)=O^+-CH_3 + H_2O \rightarrow CH_3-C(CH_3)(^+OH_2)-O-CH_3 \rightleftarrows CH_3-C(CH_3)(OH)-O-CH_3 + H^+$$

Intermediate

$$CH_3-C(CH_3)(OH)-O-CH_3 + H^+ \rightarrow CH_3-C(OH)-CH_3 + CH_3OH$$

Isopropanol

$$CH_3-OH + [O] \rightarrow HC(=O)-OH + [O] \rightarrow HO-C(=O)-OH$$

Formic Acid — Hydroxyformic Acid

$$HO-C(=O)-OH + [O] \rightarrow CO_2 + H_2O$$

$$CH_3-C(OH)-CH_3 + [O] \rightarrow CH_3-C(=O)-OH + H-C(=O)-OH$$

Acetic Acid — Formic Acid

$$CH_3-C(=O)-OH + [O] \rightarrow H-C(=O)-OH$$

Figure G.3 Potential intermediates of MTBE oxidation.

Endnotes and references

LaGrega, M. D., P. L. Buckingham, and J. C. Evans, *Hazardous Waste Management*, McGraw-Hill, New York, 1994.
Morrison, R. T. and R. N. Boyd, *Organic Chemistry*, McGraw-Hill, New York, 1998.
Perry, R. H. and D. W. Green, *Chemical Engineer's Handbook*, McGraw-Hill, New York, 1996.
Speight, J. G., *The Chemistry and Technology of Petroleum*, New York, 1990.

Appendix H

Plume geometries for subsurface concentrations of MTBE

H.1 The migration of gasoline constituents to the subsurface

The migration of gasoline containing MTBE from the surface or near surface to its encounter with the groundwater can be divided into three phases: seepage through the vadose zone, sometimes called bulk flow at the point of origin, spreading over the watertable with development of a free product layer, and accumulation in the groundwater as dissolved phase MTBE.

Figure H.1 illustrates the hypothetical case scenario of a steady-state point source leak from an underground storage tank containing gasoline and MTBE. Initially the leak moves through the high permeability sediments. The low permeability clay creates a ledge wherein the gasoline with MTBE (a nonaqueous phase liquid, or NAPL) ponds in the vadose zone. Because gasoline is a light nonaqueous phase liquid (LNAPL), free product floats on the top of the capillary fringe. In the saturated grove, a dissolved phase plume of gasoline components such as BTEX with MTBE moves with the groundwater flow direction.

In the scenario illustrated in Figures H.2, H.3, and H.4, a tanker truck transporting gasoline containing MTBE overturns, catastrophically releasing the chemicals into the environment. Figure H.3 shows a cross-section with the gasoline moving over the surface to a nearby French drain, where the contaminant moves downward into the shallow aquifer. The first clay aquitard is breached in three areas. The first, in this scenario, occurs where the clay is breached by a coarse-grained sand channel which allows vertical migration of the gasoline with MTBE into the lower middle aquifer. The second breach in the shallow aquifer occurs through an improperly abandoned irrigation well. Unfortunately, many wells (domestic, irrigation, oil, gas, geothermal, and injection) exist in the U.S. that are undocumented and therefore unregulated; these wells can create conduits to deeper aquifer zones if they were improperly constructed or abandoned. Deterioration of the annulus-sealing

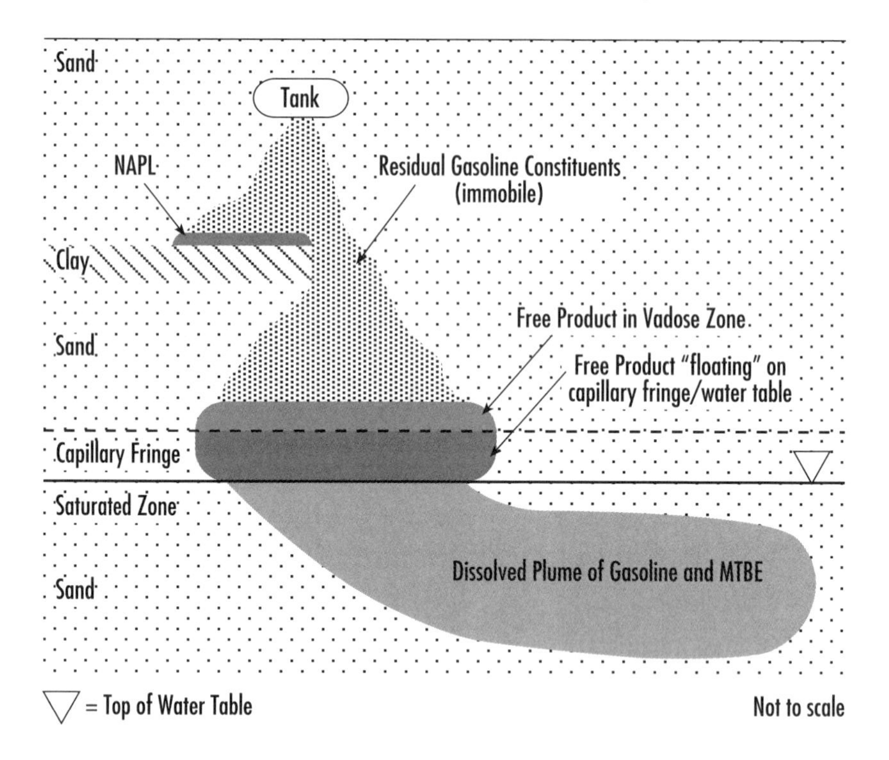

Figure H.1 Conceptual diagram of migration of gasoline containing MTBE from spill source to groundwater.

materials or the piping of these wells can result in the creation of subsurface conduits for MTBE-contaminated waters to pass through, as shown in the second breached area. In the third breached area, the water containing MTBE is diverted from the shallow aquifer into lower zones through a regional fracture system. The fracture system is quite deep and extends below the deep aquifer. By the time the spilled gasoline arrives at the deep aquifer and ultimately into the public water supply well, the MTBE has arrived first, ahead of the gasoline plume that it began with, due to the effects of adsorption and bioremediation on the gasoline components.

The hypothetical tanker truck spill shown in Figure H.4 illustrates a drawdown of a MTBE plume into the aquifer through pumping on an irrigation well. In this scenario, the MTBE plume is drawn downward into the aquifer toward the shallow irrigation well. In a related scenario, a MTBE plume rises as groundwater migrates toward a gaining lake (see Figure H.5).

H.2 MTBE forensic tools and plume geometries

Gas chromatograph analysis can be used as a forensic tool to help date MTBE releases in a relative manner. Gas chromatographs can clearly show the degradation of the BTEX compounds in relation to concentrations of MTBE.

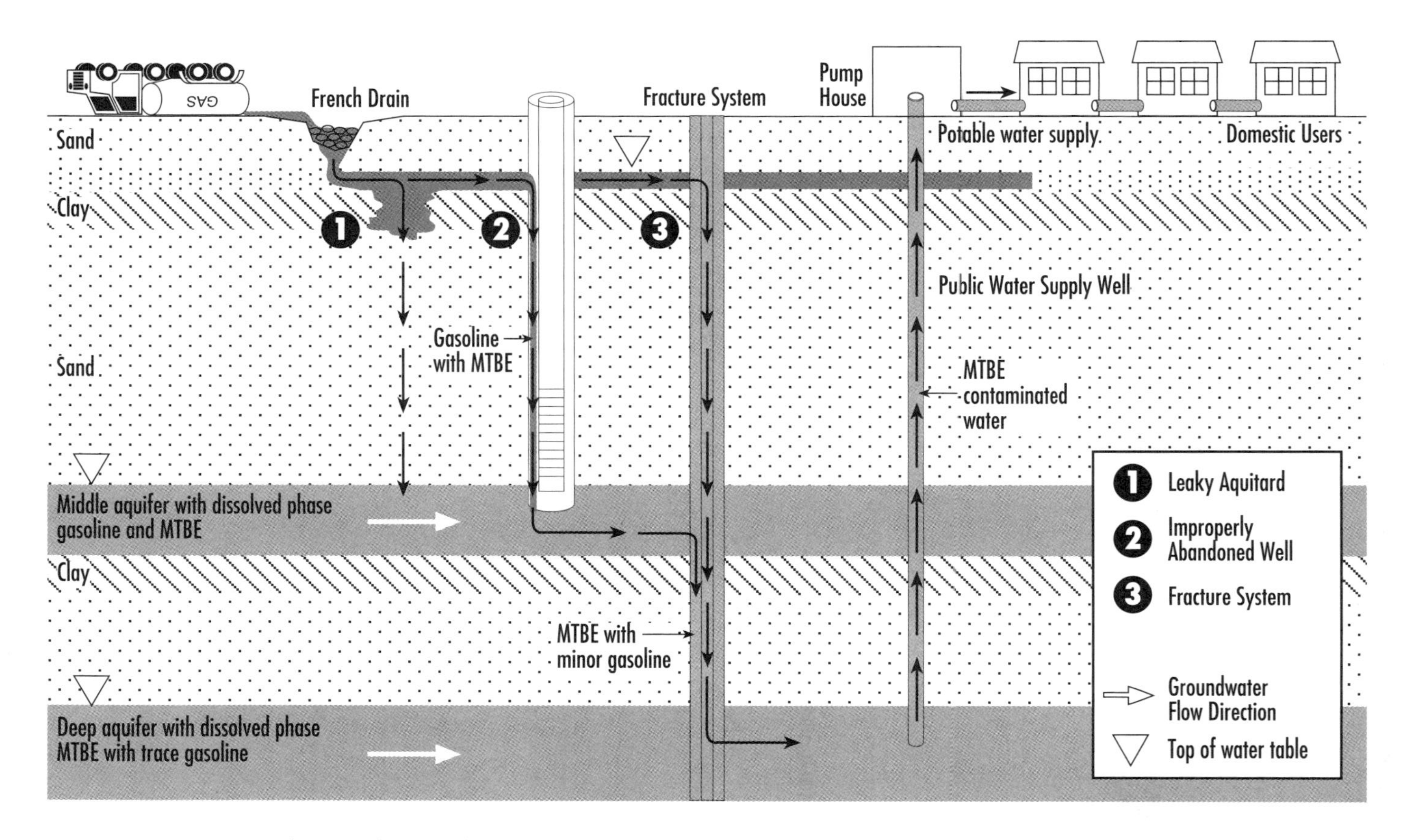

Figure H.2 Migration of MTBE from surface release to public water supply. (*Note:* not to scale.)

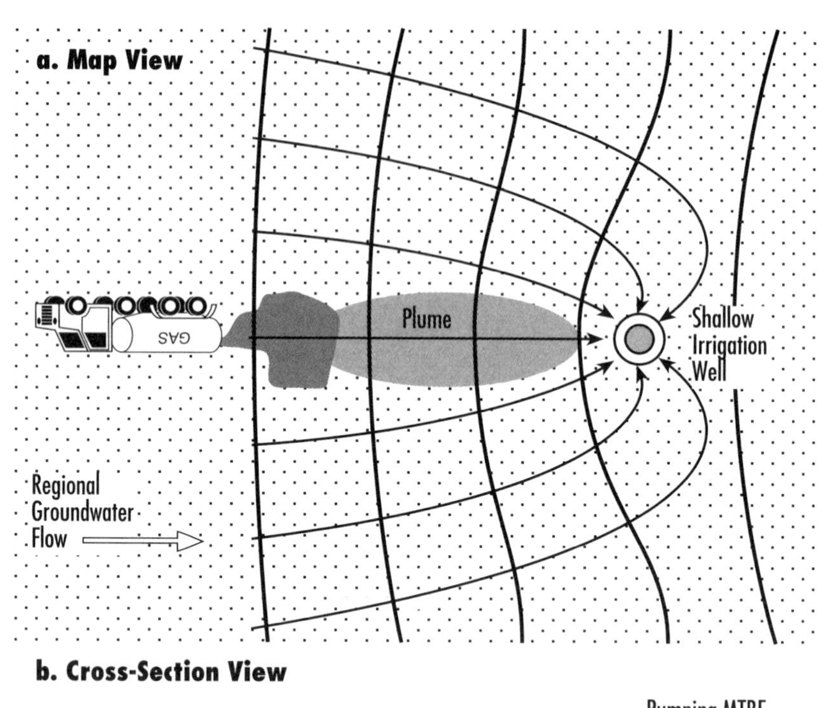

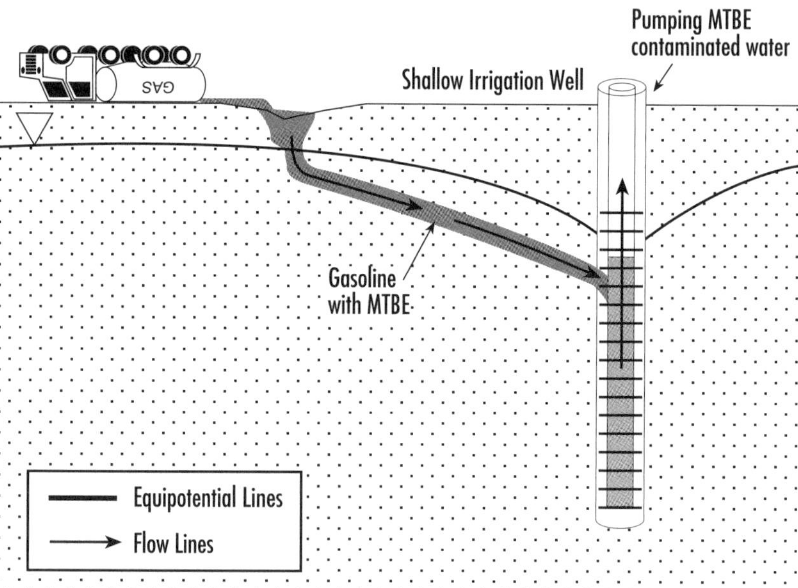

Figure H.3 MTBE plume draw-down in the aquifer through pumping.

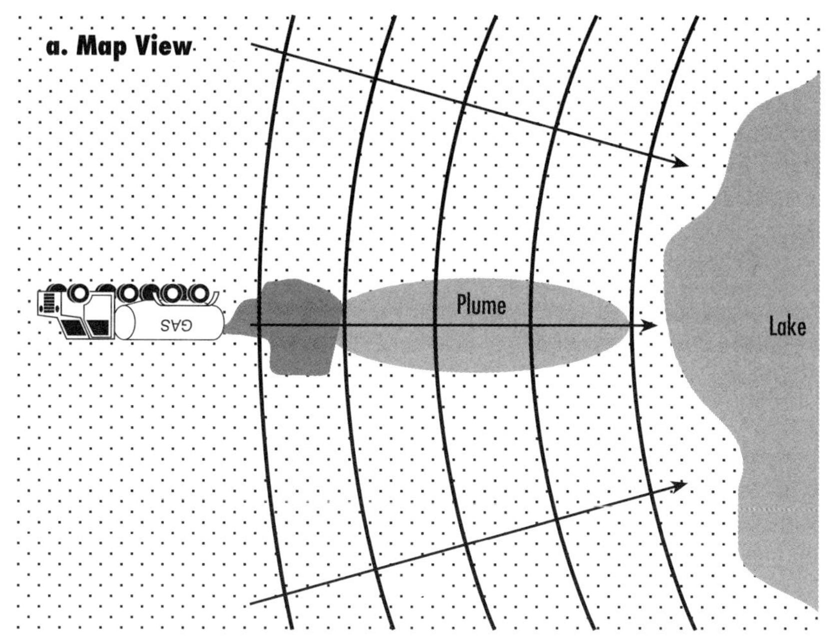

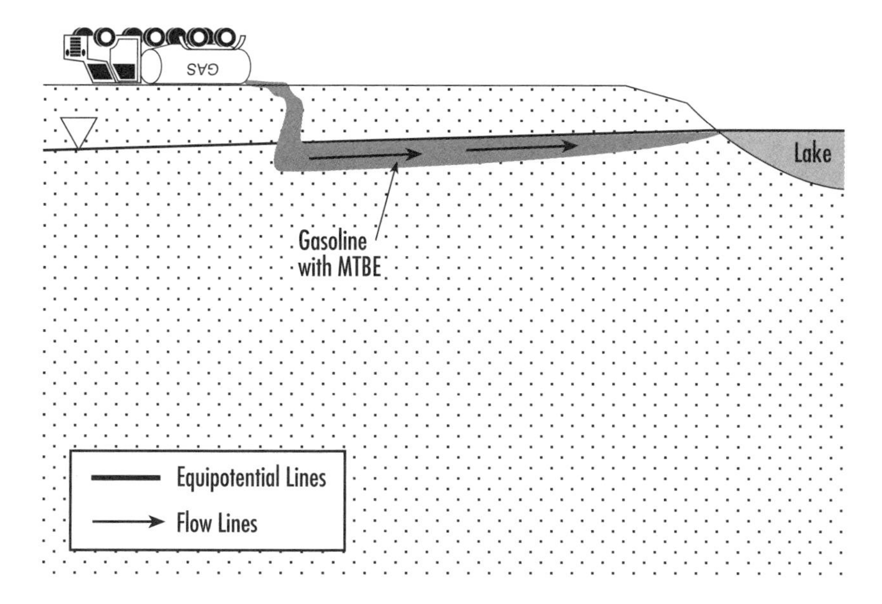

Figure H.5 Migration of MTBE toward a gaining lake.

Table H.1 Laboratory Data from Two Different Sites in California for MTBE Contaminated Groundwater (mg/L).

Number	TPHg	Benzene	Toluene	Ethylbenzene	Xylenes	MTBE	Comments
1	380	37.0	8.0	12.0	17.0	180	New Spill
2	ND	ND	ND	ND	ND	65	Old Spill

NOTES:
- TPHg = total petroleum hydrocarbons in gasoline
- ND = non-detectable concentrations
- Analysis methods used were EPA Methods 5030, Modified 8015, and 8020 or 602; and California RWQCB CSF Bay Region 7 Method GC FID 5030

SOURCE: McCampbell Analytical, 2000

Two groundwater samples (Samples 1 and 2) collected at different locations in California in April, 2000, were analyzed for petroleum constituents (see Table H.1). Sample 1 contained total petroleum hydrocarbons as gasoline (TPHg) at 380 micrograms per liter (mg/L), and benzene, toluene, ethylbenzene, and xylenes (BTEX) at levels as high as 37 mg/L. MTBE concentrations in Sample 1 were 180 mg/L. Sample 2 showed ND (non-detectable concentrations) for TPHg and the BTEX compounds. MTBE concentrations in Sample 2 were 65 mg/L. The relatively high levels of benzene in Sample 1 indicate the relatively recent nature of the leak, as benzene naturally attenuates with appropriate subsurface conditions. The lower concentrations of the TPHg and BTEX compounds in Sample 2 indicate that the release was probably older than that analyzed in Sample 1.

A gas chromatogram is the paper record from a gas chromatograph (GC) instrument. A chromatogram is plotted with time in minutes on the horizontal axis and intensity in millivolts (mV) on the vertical axis. Chromatogram peaks appear at different times and represent specific compounds that elute at their associated retention time. The height of the intensity peak or response, measured in mV, allows chemists to quantify each compound that is detected using the GC method. The higher the intensity peak, the higher the concentration of that specific compound present in a particular sample. Hypothetical gas chromatograms for Samples 1 and 2 described in the previous paragraph are shown in Figures H.6a and H.6b, respectively. The sample having the highest intensity peaks, and therefore containing more compounds is associated with Sample 1, the sample with a "fresh" chromatographic signature. The sample with a "weathered" chromatographic signature shows primarily MTBE with a few other peaks, indicating the absence of the other BTEX compounds. The interpretation that can be drawn from this information is that the BTEX compounds had naturally degraded, while the presence of MTBE persisted. One interpretation might be that Sample 2 was a chronologically older spill relative to Sample 1. Please note that interpretations may vary among chemists evaluating this type of data.

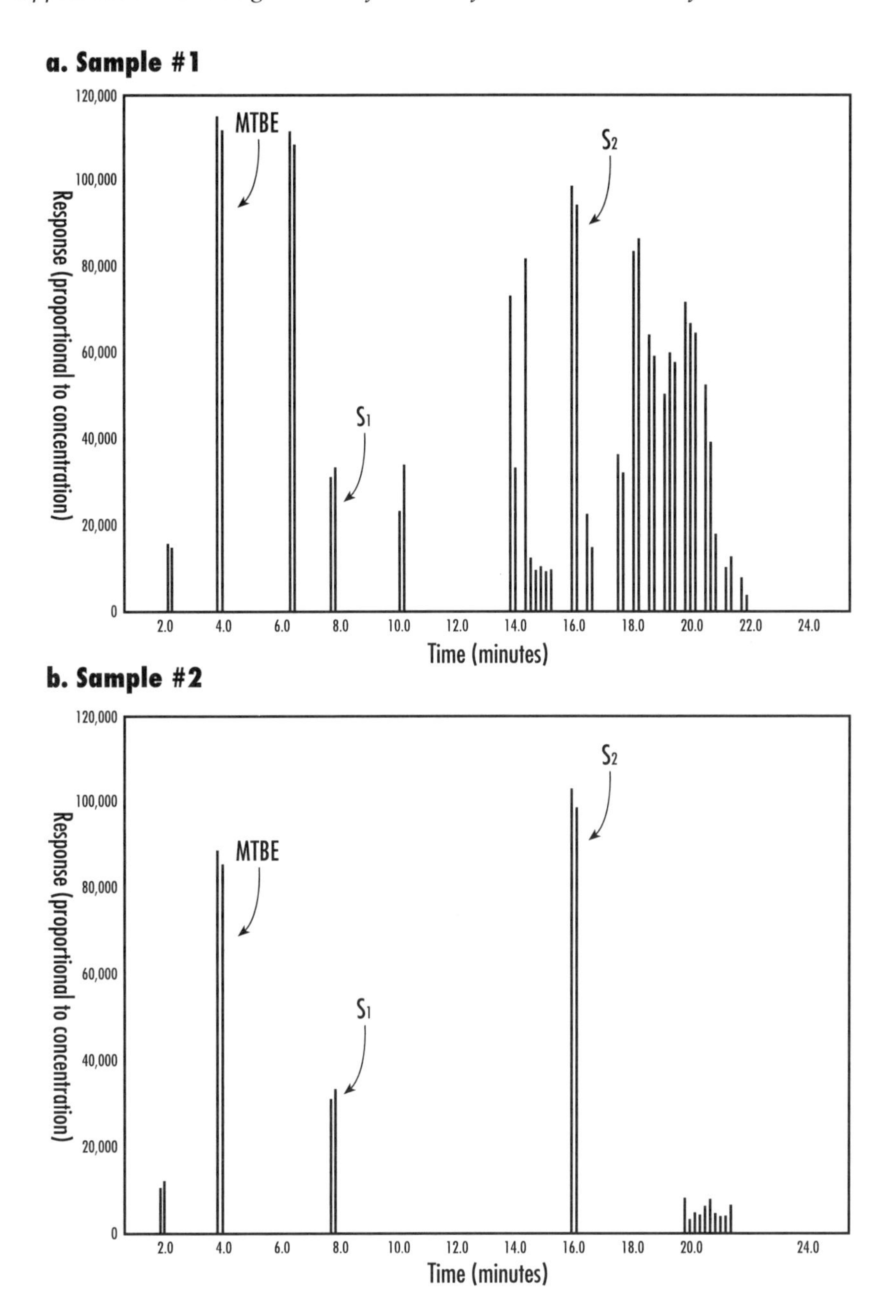

Figure H.6 Comparison of recent vs degraded gasoline samples. Notes S_1 and S_2 on each graph represent surrogates, or added compounds of known concentration used to monitor the quality of analysis.

a. Conceptual sketch of the map view of plume geometry for a point source release over three distinct release events

Tank Last Event Second Event First Event

Groundwater Flow Direction

b. Conceptual sketch of the map view of plume geometry for a point source release—steady state

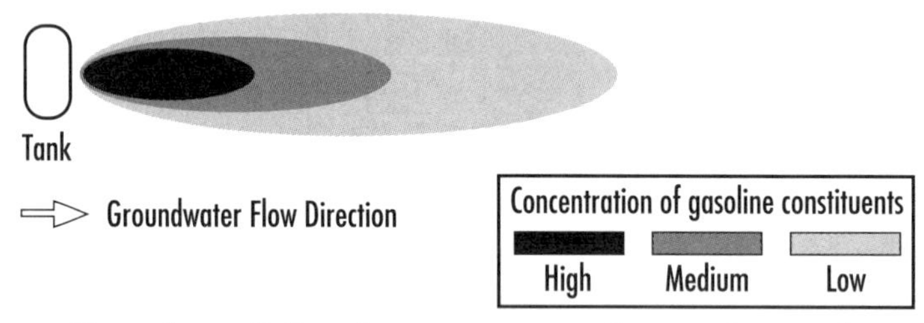

c. Map view of the plume geometry for a release of gasoline with MTBE

d. Map view of the plume geometry for a release of gasoline without MTBE followed by a later release of gasoline containing MTBE. Alternate interpretation: previously unidentified up-gradient MTBE or down-gradient gasoline plumes detected.

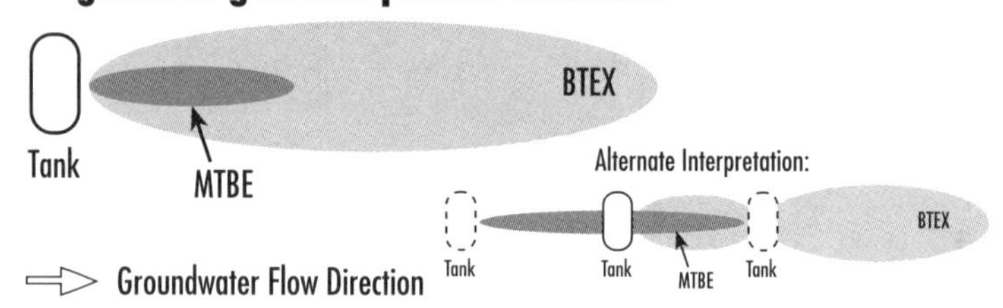

Figure H.7 Comparison of two gasoline releases.

Plume geometry can also be used as a forensic tool to determine a relative release history of a MTBE plume. A point source release over three distinct events and a steady state release are represented in Figures H.7a and b. A MTBE release that occurred concurrently with a release of associated BTEX compounds is shown in Figure H.7c. This figure shows that, after time elapses, the MTBE plume moves out in front, forming a narrow, elongated plume. The two main reasons for this plume geometry arise from MTBE's physicochemical characteristics (i.e., the BTEX compounds sorb more readily to soil than MTBE), and biodegradability (i.e., the BTEX compounds tend to biodegrade, whereas MTBE does not). A reverse case plume geometry for the release of gas without MTBE, followed by a later release of gasoline containing MTBE in the same location, is shown in Figure H.7d.

Endnotes and references

Fetter, C.F., *Contaminant Hydrogeology*, Macmillan, 2nd ed., New York, 1999.

Freeze, R.A. and Cherry, J.A., *Groundwater*, Prentice-Hall, Englewood Cliffs, New Jersey, 1979.

APPENDIX I

Toxicity of MTBE: human health risk calculations

I.1 Introduction

This appendix contains material complementary to Chapter 4, Toxicity, Health Effects, and Taste/Odor Thresholds of MTBE, in the main body of this book. Table I.1 below presents the existing regulations and advisories addressing the limits of exposure to MTBE in terms of health risk.

Table I.1 Regulations and Advisories Addressing MTBE Cancer and Noncancer Health Risks

Type of Health Risk	Advisory	Limit
Noncancer[a]	Ingestion/Dermal Sorption RfD[b]	0.03
	Inhalation RfC[c]	3 mg/m^3 (0.8ppmv)
Cancer	CPF[d]	1.8E–03 $(mg/[kg\ body\ mass]/d)^{-1}$
Inhalation	Workplace 8-h TWA[e]	100
	Acute MRL[f]	2.0
	Intermediate MRL[f]	0.7
	Chronic MRL[f]	0.7
Ingestion	Lifetime Health Advisory[g]	20
	EPA Advisory (December 1997)[h]	20 to 40
	California Public Health Goal (PHG)[h]	13
Hazardous Material	Proposition 65 (December 1998)	MTBE is not recommended for listing

[a] Office of Environmental Health and Hazard Assessment, 1999
[b] Reference Dose in mg/[kg body mass]/d
[c] Reference Concentration in mg/m^3 (parts per million by volume)
[d] Cancer Potency Factor, in $(mg/[kg\ body\ mass]/d)^{-1}$
[e] Time-weighted average (for occupational exposure), in parts per million by volume
[f] Minimal Risk Level, in parts per million by volume
[g] In micrograms per liter
[h] Maximum Concentration in Water, in micrograms per liter

I.2 Human health risk assessment: example calculation[1]

Existing data do not show that MTBE is a human carcinogen. The example calculations for the human health risk assessment are based on experimental animal (rat/mouse) carcinogenic effects. The risk of an adverse health effect from exposure to a toxic substance can be estimated by applying the following relationship:

risk a (toxicity of substance)(quantity taken into the body)

The overall process for assessing human health risk from exposure to a toxic compound consists of the following components:

1. Identification of substance of concern
2. Exposure and toxicity assessment
3. Risk calculation/estimate:

incremental cancer risk = (CPF)(intake rate)

where

CPF = Cancer Potency Factor, $(\text{mg}/[\text{kg body mass}]/\text{d})^{-1}$

intake rate = quantity of contaminant taken in per unit time, mg/[kg body mass]/d

The CPF is a regulatory value (i.e., it must usually be used in risk assessments that must be approved by a regulatory agency). It is usually derived from tests on experimental animals and assumes that there is an adverse effect at all doses, i.e., that there is no threshold for harm. To be conservative, the CPF is defined as the slope of the linear portion of the upper 95% confidence interval dose-response curve.

An incremental cancer risk of less than 1 in 1 million (less than 1.0E – 06) is almost always considered acceptable.

Incremental noncancer risk from exposure to one substance by one intake route (inhalation, ingestion, or dermal sorption) can be estimated by the following relationship:

for ingestion or dermal sorption,

Hazard Quotient = (intake rate)/(RfD)

for inhalation,

Hazard Quotient = (intake rate)/(RfC)

where

RfD = Reference Dose in mg/[kg body mass]/d

RfC = Reference Concentration in mg/[kg body mass]/d

A hazard quotient of no more than 1.0 indicates that the intake of a contaminant would result in no significant adverse effects.

The generic equation for calculating intake rate is

intake rate = [(Cim)(IRm)(FIm)(ABSF)(EF)(ED)]/[(BM)(AT)]

where

Cim = concentration of contaminant i in medium m in contact with body

IRm = intake rate of exposure medium m

FIm = fraction of intake medium m that has contaminant i

ABSF = absorption factor, fraction of i absorbed (biologically available) by the body

EF = exposure frequency, days per year (d/y)

ED = exposure duration, y

BM = body mass (kg body mass)

AT = averaging time (period over which exposure is averaged)

= ED(365 d/y) for noncancer

= (70 y)(365 d/y) for cancer

I.3 Calculation of exposure concentration for specified incremental risk (child)

The calculation for the exposure concentration, for inhalation of MTBE gas in air, that results in an estimated cancer risk of 1 in 1 million is shown below:

$$1.0E-06 = \{1.8E-03\ (\text{mg}/[\text{kg body mass}]/\text{d})^{-1}\}\{\text{intake rate}\}$$

where intake rate

$$= \{(\text{Cim, mg/m}^3)(10^{-3}\text{ mg/mg})(10\text{ m}^3\text{/d})(350\text{ d/y})(6\text{ y})\}/\{(15\text{ [kg body mass]})(70\text{ y})(365\text{ d/y})\}$$

$$= (\text{Cim, mg/m}^3)(5.479\text{E} - 05)\ (\text{m}^3\text{/mg})(\text{mg/[kg body mass]/d})$$

$$\text{Cim} = (1.0\text{E} - 06)/[(1.8\text{E} - 03)(5.479\text{E} - 05)] = 10.1\text{ mg/m}^3\ (2.8\text{ ppmv})$$

The calculation for the exposure concentration, for ingestion of MTBE in water, for a hazard quotient of 1 is shown below:

$$1 = (\text{intake rate})/(\text{RfD}) = (\text{intake rate})/(0.03\text{ mg/[kg body mass]/d})$$

where intake rate

$$= \{(\text{Cim, mg/L})(10^{-3}\text{ mg/mg})(1\text{L/d})(350\text{ d/y})(6\text{ y})\}/\{(15\text{ [kg body mass]})(6\text{ y})(365\text{ d/y})\}$$

$$= (\text{Cim, mg/L})(6.3927\text{E–05 (L/mg)(mg/[kg body mass]/d)}$$

$$\text{Cim} = 1(0.03\text{ mg/[kg body mass]/d})/\{6.3927\text{E} - 05\ (\text{L/mg})(\text{mg/[kg body mass]/d})\} = 469\text{ mg/L}$$

Thus, a child drinking MTBE-contaminated water at concentrations no greater than 469 mg/L is unlikely to experience any adverse noncancer effects. This calculated MTBE concentration for a Hazard Quotient equal to 1 is about 10 times greater than the noncancer California Public Health Concentration (PHC) of 47 mg/L for MTBE in drinking water (OEHHA 1999). This difference exists mostly because the PHC is based on a safety factor of 10,000 whereas the RfD (used in the example calculation) is typically based on a safety factor of 1000. (The example calculation above also uses different exposure factors than those used by OEHHA to calculate the PHC.)

I.4 Calculation of ingestion CPF

The calculation of the ingestion cancer potency factor is based on the California Public Health Goal (PHG) for drinking water.

The PHG for MTBE in drinking water is 13 mg/L. The following is a calculation of ingestion CPF assuming that the California Action criterion is based on a 1 in 1 million incremental cancer risk:

$$1.0E-06 = \{CPF, (mg/[kg\ body\ mass]/d)^{-1}\}\{intake\ rate\}$$

Adult (30 y of exposure) Intake Rate:

$$= \{(13\ mg/L)(10^{-3}\ mg/mg)(2\ L/d)(350\ d/y)(30\ y)\}/\{(70\ [kg\ body\ mass])(70\ y)(365\ d/y)\}$$

$$= 1.527E-04\ mg/[kg\ body\ mass]/d$$

Ingestion CPF

$$= (1.0E-06)/(1.526E-04\ mg/[kg\ body\ mass]/d)$$

$$= 6.6E-03\ (mg/[kg\ body\ mass]/d)^{-1}$$

Thus, based on the PHG of 13 mg/L for MTBE in drinking water, the calculated ingestion CPF shown above is about 3.7 times the OEHHA CPF, 1.8E – 03 $(mg/[kg\ body\ mass]/d)^{-1}$. The small difference is the result of using different exposure factors.

[1] Based primarily on

Agency for Toxic Substances and Disease Registry, The Toxicological Profile for Methyl *tert*-Butyl Ether, U.S. Department of Health and Human Services, Agency for Toxic Substances and Disease Registry, 1996.

Office of Environmental Health Hazard Assessment, Public Health Goal for Methyl *Tertiary*-Butyl (MTBE) in Drinking Water, March 1999.

Appendix J

MTBE web sites

Below is a limited selection of informative web sites about MTBE. Because California agencies and groups are so involved with MTBE research and policy development, it is not surprising that a large number of MTBE web sites are associated with California organizations.

FEDERAL GOVERNMENT

United States Environmental Protection Agency (U.S. EPA) has a variety of web sites:

http://www.epa.gov/swerust1/mtbe/index.htm
http://www.epa.gov/swerust1/mtbe/mtbemap.htm
http://www.epa.gov/OST/Tools/dwstds.html
http://www.epa.gov/oms/regs/fuels/oxy-area.pdf
http://www.epa.gov/oms/consumer/fuels/oxypanel/blueribb.htm
http://www.epa.gov/opptintr/chemfact/s-mtbe.txt
http://www.epa.gov/otaq/consumer/fuels/oxypanel/blueribb.htm

Comments: Mostly unbiased information from the EPA

United States Geological Survey (USGS) has been studying MTBE for several years. The U.S. Geological Survey has a MTBE website and maintains a bibliography of scientific literature regarding MTBE and other fuel oxygenates. Some of the early work on water quality was performed by the USGS on urban areas.

http://water.wr.usgs.gov/mtbe/
http://wwwsd.cr.usgs.gov/nawqa/vocns/mtbe/bib/
http://wwwsd.cr.usgs.gov/nawqa/vocns/mtbe/bib/key.html

Comments: Fate and transport of MTBE, and other articles and case studies make these interesting MTBE web sites.

U.S. NAVY

Naval Facilities Engineering Service Center (NFESC), Pt. Hueneme, California

http://www.nfesc.navy.mil

Comments: Technology demonstration web site featuring many MTBE technologies. When a technology has proven to be successful, documents will be available on the web page.

NATIONAL LABORATORIES

Lawerence Livermore National Laboratory, Livermore, California

http://www.erd.llnl.gov/mtbe
http://geotracker.llnl.gov/

Comments: This national lab has provided a database of sites with MTBE, plume lengths and other data compilations.

ACADEMIC INSTITUTIONS

University of California at Davis

http://trg.ucdavis.edu/clients/trg/research/mtbe.html
http://tsrtp.ucdavis,edu/mtbe

Comments: A UC Davis web site showing research on the source, fate, and transport of the controversial gasoline additive MTBE in Lake Tahoe, California MTBE research at UC Davis is thought provoking.

WATER AGENCIES

Association of California Water Agencies (ACWA): The ACWA is a California organization whose 437 public water agency members collectively manage and deliver 90% of the urban and agricultural water used in the state.

http://www.acwanet.com/legislation/regulatory/mtbe2.html

Comments: Links and regulatory comments from water agencies.

Santa Clara Valley Water District (San Jose area, California): A leading regulatory and water agency in Silicon Valley, California.

http://www.scvwd.dst.ca.us/wtrqual/ustmtbe.htm
http://www.swrcb.ca.gov/baydelta/mtbe_finaldraft.doc

Comments: Practical information from a water district with many MTBE sites. Prioritization method developed by SCVWD.

www.calepa.ca.gov/programs/mtbe

Comments: State of California's response to MTBE

TRADE GROUPS AND INSTITUTES

American Petroleum Institute (API): The American Petroleum Institute has an oxygenate bibliography, a list of online MTBE reports, and links to other sites:

http://www.api.org/ehs/mtbelink.htm
http://www.api.org/ehs/apibib.htm
http://www.api.org/ehs/othrmtbe.htm

Comments: Good information on MTBE from an oil industry-based institute.

Oxygenated Fuels Association (OFA)

http://ofa.net

Comments: A trade group representing the MTBE manufacturers, this web page provides several interesting viewpoints to consider, albeit highly biased toward selling their product, MTBE.

VENDORS

There are several vendors who claim success with MTBE. The authors do not recommend or endorse particular vendors.

Comments on vendors' web sites: For information on in-situ remediation methods of MTBE at the time of this publication, caution and evaluation of vendors' claims and data are warranted. Decreases in MTBE concentrations in groundwater can be due to remediation technologies, as well as other factors such as seasonal variations in hydrogeologic conditions and water levels, dilution, dispersion, poor sampling programs, limited data, or other factors.

Appendix K

Summary of MTBE remediation technologies

Remediation of MTBE is still more challenging than other gasoline related compounds. Site-specific subsurface geologic conditions including heterogeneity of soil or sediments, permeability, porosity, and other site-specific parameters greatly affect remediation success of MTBE. Nonetheless, remediation potential of various soil and groundwater technologies can be compared with the physio-chemical characteristics of MTBE. Without endorsing specific products, manufacturers, or services, the list below summarizes selected methods that have been used in the field or laboratory.

K.1 Summary of soil characteristics

Remediation of MTBE impacted soil must take into account the adsorption coefficent of MTBE. The soil sorption or distribution coefficient (Kd) is also known as the adsorption coefficient; it is a chemical's ability to bind to soil particles. The greater the Kd, the greater the chemical's ability to sorb to soil. The Kd is defined in the equation:

$$Kd = Cs/Ce$$

where

Cs = concentration adsorbed on soil surfaces in mg/kg (μg/g)
Ce = concentration in water in mg/L (μg/mL).

A chemical's Kd may also be calculated from the soil/sediment organic carbon coefficient or Koc using the equation

$$Kd = Koc \times foc$$

where

Koc = the soil/sediment organic carbon to water coefficient (mL/g)
foc = the fraction of organic carbon in soil (e.g., 0.01)

Values published are generally published in mL/g.

Dragun (1988) and Olsen and Davis (1990), defined hydrocarbon mobility by the Kd and Koc. Howard and others (1993) give a Koc of 11.2 for MTBE. Therefore, for a fairly typical silty soil with a foc of 5% (0.05), the Kd would be

$$11.2 \text{ x } 0.05 = 0.56$$

The Mobility Index is summarized in Table K.1. In a sandy soil with much less organic carbon, the Koc might be 1% (0.01) which would give the following Kd:

$$1.2 \times 0.01 = 0.112$$

Therefore, based on the soil sorption coefficient, MTBE is mobile (Class IV) and based on the Koc, it is very mobile. Owing to the mobile nature of MTBE, most remediation technologies (both in situ and aboveground) must account for a longer treatment retention time than for other gasoline compounds.

Table K.1 Mobility Index Based on Kd and Koc

Class	Type	Kd	Koc
I	Immobile	> 10	> 2000
II	Low mobility	2 – 10	500 – 2000
III	Intermediate	0.5 – 2	150 – 500
IV	Mobile	0.1 – 0.5	50 – 150
V	Very Mobile	< 0.1	< 50

K.2 Summary of soil remediation for MTBE

K.2.1 Conventional soil remediation (soil)

Conventional remediation of soil by excavation is reasonably predictable and successful to depths of about 20 to 30 ft, the practical reach limit of a range of equipment such as backhoes and excavators. Transportation and disposal costs for the MTBE- impacted soils can be significant. After soil is excavated, on-site treatment of MTBE-contaminated soil may be performed using low temperature thermal desorption, providing soil characteristics are favorable. The large expense of excavation projects as well as major site disruption can be an important factor for choosing alternative in-situ methods.

K.2.2 Soil vapor extraction (soil)

Soil Vapor Extraction (SVE) uses a vacuum to draw the contaminants in vapor form from the void spaces through a network of vadose zone wells to the surface for destruction. With high vapor pressure, MTBE is amenable to SVE.

K.2.3 Aerobic biodegration (soil and groundwater)

Aerobic biodegradation within the vadose is challenging, since MTBE is considered recalcitrant. Consequently, these methods are generally less successful than other soil remediation methods for MTBE.

K.2.4 Advanced oxidation (soil and groundwater)

Oxidation chemicals such as hydrogen peroxide have been documented to oxidize MTBE. Systems for oxidant delivery include wells, trenches, or filter galleries. Another method of oxidant delivery is jetting, which uses high-pressure injection through a lance driven into the subsurface on closely spaced centers. Regardless of the delivery system, in-situ applications of oxidation chemicals are not commonly used in the environmental field at present due to concerns with employee safety and risks in handling high concentrations of strong oxidizers such as hydrogen peroxide. Important variables for in-situ oxidation of soil include pH, alkalinity, total organic carbon, iron content, as well as the permeability and porosity of the soil.

The relative reactivities of the various oxidants are listed in Table K.2.

Table K.2 Relative Oxidizing Power of Selected Chemical Oxidants

Reative Species	**Relative Oxidizing Power (Cl_2 = 1.0)**
Fluorine	2.23
Hydroxyl Radical (Fenton's reagent)	2.06
Hydrogen Peroxide	1.31
Permanganate	1.24
Chlorine Dioxide	1.15
Chlorine	1.00
Bromine	0.80
Iodine	0.54

K.2.5 MTBE in-situ groundwater remediation technologies

K.2.5.1 Passive in-situ remediation treatment wall

A passive permeable treatment wall constructed into the Groundwater allows for installation of a variety MTBE treatment media (Figure K.1). Treat-

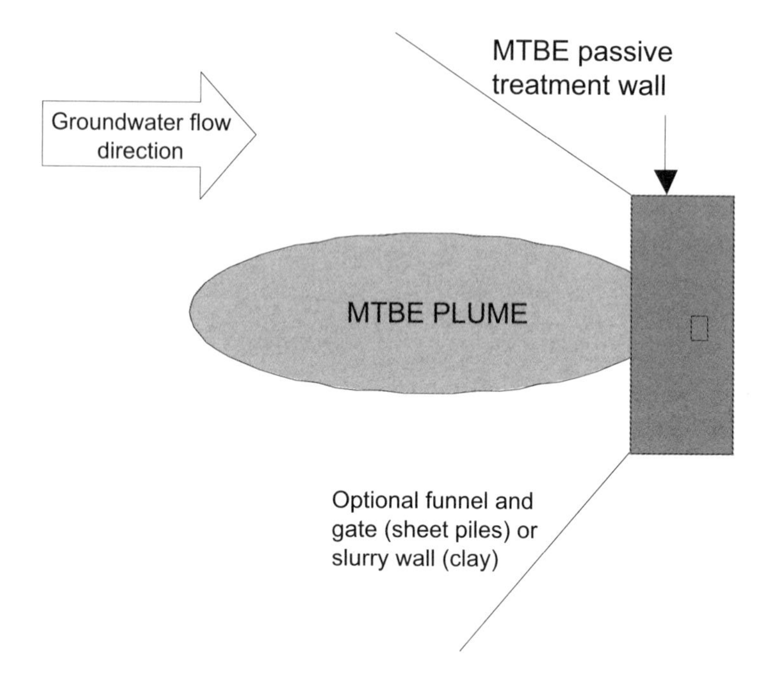

Figure K.1 Plan view, sketch of passive funnel and gate treatment wall.

ment media may include adsorbents such as granular activated carbon (GAC), granular organic polymers, and synthetic resins. Although originally designed as aboveground treatment technologies, injecting or placing these adsorbent materials in the subsurface where Groundwater flow can pass through a permeable treatment wall, can enhance in-situ remediation methods of MTBE. Adsorbent media, such as carbon, may require maintenance and replacement when MTBE breakthrough occurs. Subsurface chemistry can be augmented with organic acid catalysts, and chemically treated iron particles for Fenton's Chemistry.

Passive bioremediation treatment media may include, but are not limited to products that release oxygen, nutrients, or other products. Downgradient monitoring of passive treatment walls is essential to long-term success. Treatment walls work well if groundwater flow direction is relatively constant. An optional funnel and gate system using sheet piles or bentonite slurry walls can be used to contain and direct the MTBE-impacted water into the permeable treatment zone if groundwater flow directions migrate during the hydrologic cycle.

In-situ projects generally require three stages: bench or laboratory testing, pilot testing, and final implementation. Design considerations include detailed geochemistry evaluations and bench tests for the interaction of the MTBE-impacted Groundwater and the treatment media. This phase is important to verify that the treatment media or treatment chemicals will work

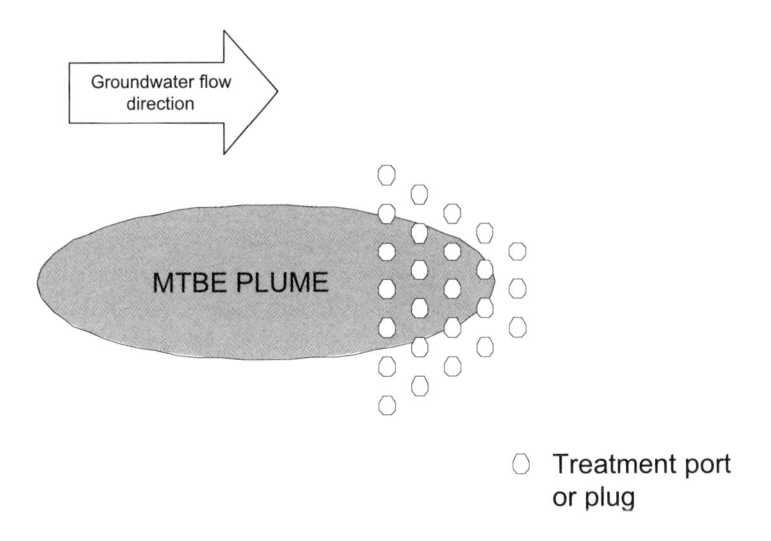

Figure K.2 Plan view, treatment ports or plugs for in-situ remediation by jetting and plugging.

with MTBE in the site-specific subsurface conditions. For example, if hydrogen peroxide is to be used to oxidize MTBE, certain parameters are needed: alkalinity, total organic carbon (TOC), iron content, and pH are needed for soil and alkalinity, acidity, and pH for the water. Hydrogen peroxide works best with lower pH ranges, and acid treatments may be recommended for sites with pH readings over 7. At this stage, the hydrogeologic parameters such as permeability, porosity, and preferred flow paths of the site are evaluated.

During the pilot test of an in-situ project, a small, but representative area is treated and evaluated. Based on the results of the pilot test, a full scale remediation is designed and implemented. The cost for treatment is reasonable at groundwater depths between 5 and 25 ft. Cost of treatment increases greatly at depths exceeding 25 ft.

Jetting is an in-situ remediation technology that uses numerous treatment ports on closely spaced centers (3 to 5 ft centers) to deliver a variety of chemical oxidants (hydrogen peroxide, potassium permanganate or others) to oxidize MTBE (Figure K.2). Hydrogen peroxide reacts quickly (within minutes to hours) while potassium permangante reacts more slowly (hours to days). Close spacing of the injection ports works well for lower permeability zones. Jetting typically injects chemicals into the ground at pressures of 3000 to 5000 psi with each port receiving up to 100 gallons of liquid. Jetting can be used for in-situ bioremediation to introduce chemicals such as enzymes, nutrients, or other amendments.

K.2.5.2 Bioremediation of MTBE in groundwater

Enhanced aerobic bioremediation by plugging, uses a similar treatment configuration as shown in Figure K.2. Magnesium peroxide is injected (up to 1000 psi) into the subsurface as a slurry. Upon curing, the magnesium peroxide slurry turns into solid vertical plugs, approximately 1 to 2 ins in diameter. When wet, these plugs release oxygen over a period of weeks to months. Injection of 30 to 50 lb of magnesium peroxide per borehole is not uncommon, depending on the subsurface permeability and porosity. Aerobic bioremediation of MTBE-impacted groundwater using magnesium peroxide has been performed in the laboratory and in the field using specially enhanced consortia of naturally occurring microbes. However the results are mixed and inconclusive at this time. Consistent successful results are needed before engineered bioremediation of MTBE can be used on a widespread basis.

In both jetting and plugging, after the ports or boreholes are stabilized, they are sealed with a neat cement or bentonite grout to the surface to prevent surface infiltration into the subsurface.

K.2.5.3 Pressure injection using a well network

Propane-oxidizing bacteria can be used to biodegrade MTBE using either jetting technologies or wells. A sparge system has been developed to inject propane and oxygen into the MTBE-impacted aquifer. The propane and oxygen encourage the growth of the propane-oxidizing bacteria and their production of the enzyme propane monooxygenase that catalyzes the destruction of MTBE.

A network of wells (Figure K.3) can be used to inject hydrogen peroxide under high pressure. The pressure is used to insure that the peroxide spreads out relatively evenly in the subsurface. Injection through wells works best with highly permeable aquifers.

K.2.5.4 Air sparging

Using a relatively closely spaced well network, air sparging works by forcing air through the Groundwater zone, removing the contaminants by volatilization and, if applicable, aerobic microbial degradation. Based on physiochemical characteristics of MTBE, air sparging should be somewhat effective as a remedial option for MTBE.

K.2.6 Extraction technologies

Commonly called "pump and treat," Groundwater containing MTBE can be pumped from an extraction well, trench, horizontal well network to the surface for treatment (Figure K.3). Surface treatment has been proven to be successful for MTBE.

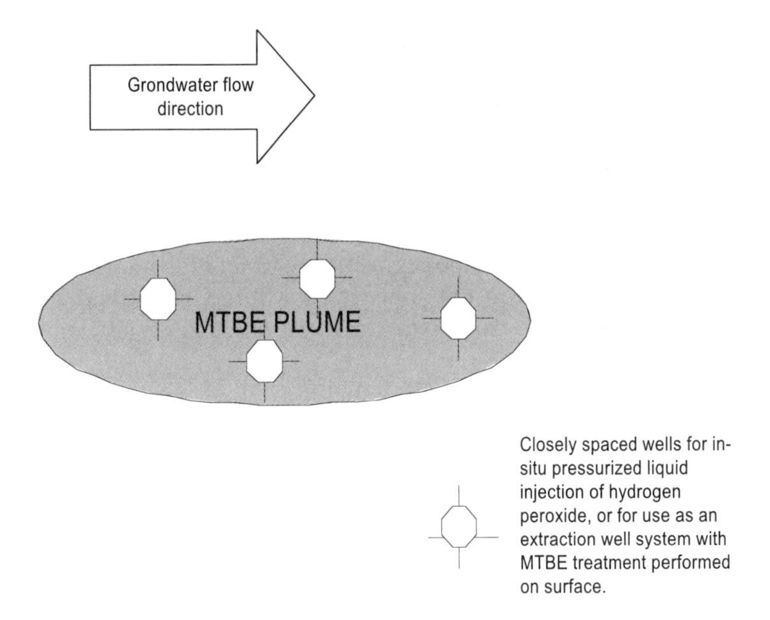

Figure K.3 Well network used for pressure injection of liquids or extraction of MTBE-contaminated water to be treated aboveground.

K.2.7 Advanced oxidation technology

Some of the promising aboveground treatment technologies include advanced oxidation processes: hydrogen peroxide/ozone systems and ultraviolet oxidation/ozone (UV/ozone) systems. These systems rely on the hydroxyl radical (OH) to chemically oxidize MTBE into the ultimate end products of carbon dioxide and water. Ozone is an oxidant gas generated onsite. Variations in the systems include catalytic and non-catalytic options. Although there is more processing and maintenance with the catalytic options, these systems tend to treat MTBE faster than non-catalyzed processes. Advanced oxidation technology is an economical treatment of high concentrations of MTBE, not trace levels.

K.2.8 Adsorption technologies

Adsorption technologies include granular activated carbon (GAC) and synthetic resin sorbents. GAC comes in a variety of particle sizes and grades. MTBE's capacity for each grade of carbon should be evaluated in the laboratory or during the pilot study prior to a full-scale remediation program.

Synthetic resins are regenerable on site and have greater capacity and generally lower life-cycle costs than the equivalent weight of carbon in equivalent applications. Higher capital costs are associated with synthetic resins.

A granular organic polymer with high absorption of MTBE from water is non-regenerable and relatively low performance.

MTBE-contaminated groundwater with concentrations higher than 100 parts per million can be treated aboveground using granular organic polymers; however, at lower concentrations, and for short-term applications, granular activated carbon is better. For lower concentrations (<100 ppm) and for long-term applications, synthetic resins will be the most economical choice. Although designed as aboveground treatment technologies, injecting or placing these materials in the subsurface where Groundwater flow can pass through a permeable treatment wall, can enhance in-situ remediation methods of MTBE.

Other aboveground technologies include air stripping. Although air stripping is not considered to be practical due to the low Henry's law constant of MTBE, the technology has been used in removal of MTBE from Groundwater.

K.2.9 MTBE risk-based evaluation

Although not a treatment technology, risked-based assessments are useful in characterizing risks to potential human and wildlife receptors. The risk analysis evaluates MTBE sites and prioritizes them based on risk to human health or exposure to Groundwater or surface waters and wildlife. The prioritization can be used for funding or determining which sites are likely to impact drinking water supplies in the near future.

Endnotes and references

Howard, P.H., Michalenko, E.M., Basu, D.K., Sage, G.W., Meylan, W.M., Beauman, J.A., Jarvis, W.F., and Gray, D.A., Eds., *Handbook of Environmental Fate and Exposure Data for Organic Chemicals*, Volume IV, *Solvents*, Lewis Publ., Chelsea, MI, 578 p., 1993.

Dragun, J., *The Soil Chemistry of Hazardous Materials*, Hazardous Materials Control Research Institute, Silver Spring, MD, 458 p., 1988.

Olsen, R.L. and Davis, A., Predicting the fate and transport of organic compounds in groundwater: *Hazardous Materials Control*, v.3, n.3, pp. 39-64, 1990.

Testa, S.M. and Winegardner, D.L., *Restoration of Contaminated Aquifers, Petroleum Hydrocarbons and Organic Compounds*, Lewis Publ., Boca Raton, FL, 446 pp., 2000.

Bibliography and reading list

Ainsworth, S., Oxygenates seen as hot market by industry, *Chemical & Engineering News*, v. 70, no. 19, p. 29–30, 1992.

Alberta Research Council, Composition of Canadian summer and winter gasolines, Canadian Petroleum Products Institute Report No. 94–5, p. A-65–A-118, B-67–B-118, June 1994.

Alexander, J.E., Ferber, E.P., and Stahl, W.M., Avoid leaks from reformulated fuels—Choose an elastomeric sealing material according to the type and concentration of oxygenate (ether and/or alcohol) added to the fuel, *Fuel Reformulation*, p. 42–46, March/April 1994.

Allan, R.D. and Parmele, C.S., Treatment technology for removal of dissolved gasoline components from groundwater, in National Symposium on Aquifer Restoration and Groundwater Monitoring, Columbus, Ohio, May 25–27 1983.

Allen, M. and Grande, D., Reformulated gasoline air monitoring study, Madison, State of Wisconsin Department of Natural Resources, Bureau of Natural Resources, American Petroleum Institute Publication AM-175-95, 1995.

American Conference of Governmental Industrial Hygienists, 1994–1995, Threshold limit values for chemical substances and physical agents and biological exposure indices, Cincinnati, Ohio, p. 100, 1995.

American Petroleum Institute, Alcohols and ethers—A technical assessment of their application as fuels and fuel components (2nd ed.), Washington, D.C., Refining Department, American Petroleum Institute Publication No. 4261, 1988.

American Petroleum Institute, A compilation of field-collected cost and treatment effectiveness data for the removal of dissolved gasoline components from groundwater, Washington, D.C., Health and Sciences Department, American Petroleum Institute Publication No. 4525, 1990.

American Petroleum Institute, Field evaluation of biological and non-biological treatment technologies to remove MTBE/oxygenates from petroleum product terminal wastewaters, Washington, D.C., Health and Sciences Department, American Petroleum Institute Publication No. 4655, 1997.

American Petroleum Institute, Field studies of BTEX and MTBE intrinsic bioremediation, Washington, D.C., Health and Sciences Department, American Petroleum Institute Publication No. 4654, 1997.

American Petroleum Institute, Delineation and characterization of the Borden MTBE plume—An evaluation of eight years of natural attenuation processes, Washington, D.C., Health and Sciences Department, American Petroleum Institute Publication No. 4668, 1998.

American Society for Testing and Materials, Standard test method for determination of C1 to C4 alcohols and MTBE in gasoline by gas chromatography, in American Society for Testing and Materials, annual book of ASTM standards, Philadelphia, PA, p. 631–635, 1988.

Ames, T.T. and Grulke, E.A., Group contribution method for predicting equilibria of non-ionic organic compounds between soil organic matter and water, *Environmental Science & Technology*, v. 29, no. 9, p. 2273–2279, 1995.

Anderson, E.V., Health studies indicate MTBE is safe gasoline additive, *Chemical & Engineering News*, v. 71, no. 38, p. 9–18, 1993.

Anderson, H.A., Hanrahan, L., Goldring, J., and Delaney, B., An investigation of health concerns attributed to reformulated gasoline use in southeastern Wisconsin, Wisconsin Department of Health and Social Services, Division of Health, Bureau of Public Health, Section of Environmental Epidemiology and Prevention, Final Report, 1995.

Anderson, L.G., Wolfe, P., Barrell, R.A., and Lanning, J.A., The effects of oxygenated fuels on the atmospheric concentrations of carbon monoxide and aldehydes in Colorado, in Sterrett, F.S., Ed., *Alternative Fuels and the Environment*, Ann Arbor, Mich., Lewis Publishers, Inc., p. 75–103, 1994.

Anderson, L.G., Wolfe, P., and Wilkes, E.B., Effects and effectiveness of using oxygenated fuels in the Denver metropolitan area, in American Chemical Society Division of Environmental Chemistry pre-prints of papers, 213th, San Francisco, California, ACS, v. 37, p. 381-383, 1997.

Angle, C.R., If the tap water smells foul, think MTBE, *Journal of the American Medical Association*, v. 266, no. 21, p. 2965–2966, 1991.

Baehr, A.L., Baker, R.J., and Lahvis, M.A., Transport of methyl *tert*-butyl ether across the water table to the unsaturated zone at a gasoline-spill site in Beaufort, S.C., in American Chemical Society Division of Environmental Chemistry pre-prints of papers, 213th, San Francisco, California, ACS, v. 37, p. 417–418, 1997.

Baehr, A.L., Stackelberg, P.E., Baker, R.J., Kauffman, L.J., Hopple, J.A., and Ayers, M.A., Design of a sampling network to determine the occurrence and movement of methyl *tert*-butyl ether and other organic compounds through the urban hydrologic cycle, in American Chemical Society Division of Environmental Chemistry pre-prints of papers, 213th, San Francisco, California ACS, v. 37, p. 400–401, 1997.

Barker, J.F., Hubbard, C.E., and Lemon, L.A., The influence of methanol and MTBE on the fate and persistence of monoaromatic hydrocarbons in groundwater, *Groundwater Management*, v. 4, p. 113–127, 1990.

Barker, J. F., Schirmer, M., and Hubbard, C.E., The longer term fate of MTBE in the Borden Aquifer, in Stanley, A., Ed., NWWA/API Petroleum Hydrocarbons and Organic Chemicals in Groundwater—Prevention, Detection, and Remediation Conference, Houston, Tex., Nov. 13–15, 1996.

Barker, J. F., Schirmer, M., Butler, B.J. and Church, C.D., Fate and transport of MTBE in groundwater — results of a controlled field experiment in light of other experience, in The Southwest Focused Groundwater Conference—discussing the Issue of MTBE and Perchlorate in Groundwater, Anaheim, California, June 3–4, 1998.

Barreto, R.D., Gray, K.A., and Anders, K., Photocatalytic degradation of methyl-*tert*-butyl ether in TiO2 slurries—a proposed reaction scheme, *Water Resources*, v. 29, no. 5, p. 1243–1248, 1995.

Baur, C., Kim, B., Jenkins, P.E., and Cho, Y., Performance analysis of SI engine with ethyl tertiary butyl ether (ETBE) as a blending component in motor gasoline and comparison with other blending components, in Nelson, P.A., Schertz, W.W., and Till, R.H., Eds., Intersociety Energy Conversion Engineering Conference, 25th, Proceedings, v. 4, New York, American Institute of Chemical Engineers, p. 337–342, 1990.

Beckenbach, E.H. and Happel, A.M., Methyl *tertiary*-butyl ether plume evolution at California LUFT sites, in The Southwest Focused Groundwater Conference—Discussing the Issue of MTBE and Perchlorate in Groundwater, Anaheim, California, June 3-4, Proceedings, Anaheim, California, National Groundwater Association, p.15, 1998.

Begley, R. and Rotman, D., Health complaints fuel federal concern over MTBE, *Chemical Week*, v. 152, no. 10, p. 7, 1993.

Begley, R., MTBE high demand time looms as health questions linger, *Chemical Week*, v. 155, no. 8, p. 13, 1994.

Belpoggi, F., Soffritti, M., and Maltoni, C., Methyl *tertiary*-butyl ether (MTBE)—a gasoline additive—causes testicular and lymphohaematopoeitic cancers in rats, *Toxicology and Industrial Health*, v. 11, no. 2, p. 1–31, 1995.

Bennett, P.J. and Kerr, J.A., Kinetics of the reactions of hydroxyl radicals with aliphatic ethers studied under simulated atmospheric conditions—Temperature dependencies of the rate coefficients, *Journal of Atmospheric Chemistry*, v. 10, no. 1-2, p. 27–38, 1990.

Bhattacharya, A.K. and Boulanger, E.M., Organic carbonates as potential components of oxygenated gasoline (abs.), in American Chemical Society Division of Environmental Chemistry pre-prints of papers, 208th, Washington, D.C., ACS, v. 34, no. 2, p. 471–473, 1994.

Bianchi, A. and Varney, M.S., Analysis of methyl *tert*-butyl ether and 1,2-dihaloethanes in estuarine water and sediments using purge-and-trap/gas-chromatography, *Journal of High Resolution Chromatography*, v. 12, no. 3, p. 184–186, 1989.

Biles, R.W., Schroeder, R.E., and Holdsworth, C.E., Methyl *tertiary*-butyl ether inhalation in rats—A single generation reproduction study, *Toxicology and Industrial Health*, v. 3, no. 4, p. 519–534, 1987.

Bobro, C.H., Karas, L.J., Leaseburge, C.D., and Skahan, D.J., Decreased benzene evaporative emissions from an oxygenated fuel, pre-printed papers, in American Chemical Society Division of Fuel Chemistry pre-prints of papers, ACS, v. 39, no. 2, p. 305–309, 1994.

Boggess, K., Analysis of human blood specimens for methyl *tertiary*-butyl ether (MTBE) and tertiary butyl alcohol (TBA), Kansas City, Mo., Midwest Research Institute, MRI Project No. 3454, 1994.

Boughton, C.J. and Lico, M.S., 1998, Volatile organic compounds in Lake Tahoe, Nevada and California, July-September , U.S. Geological Survey Fact Sheet FS-055-98, 1997.

Bolton, J.R., Safarzadeh-Amiri, A., Cater, S.R., Dussert, B., Stefan, M., and Mack, J., Mechanism and efficiency of the degradation of MTBE in contaminated groundwater by the UV/H202 Process, in The Southwest Focused Groundwater Conference—Discussing the Issue of MTBE and Perchlorate in Groundwater, Anaheim, California, June 3-4, Anaheim, California, National Groundwater Association, p. 36–39, 1998.

Bonin, M.A., Ashley, D.L., Cardinali, F.L., McCraw, J.M., Moolenaar, R.L., Hefflin, B.J., Etzel, R.A., and Wooten, J.V., Measurement of methyl *tert*-butyl ether and butyl alcohols in whole human blood by purge-and-trap gas chromatography-mass spectrometry (abs.), in American Chemical Society Division of Environmental Chemistry pre-prints of papers, 206th, Chicago, Ill., ACS, v. 33, no. 2, p. 21–24, 1993.

Bonin, M.A., Ashley, D.L., Cardinali, F.L., McCraw, J.M., and Wooten, J.V., Measurement of methyl *tert*-butyl ether and tert-butyl alcohol in human blood and urine by purge-and-trap gas chromatography-mass spectrometry using an isotope-dilution method (abs.), in American Chemical Society Division of Environmental Chemistry pre-prints of papers, 208th, Washington, D.C., ACS, v. 34, no. 2, p. 153–155, 1994.

Bonin, M.A., Ashley, D.L., Cardinali, F.L., McCraw, J.M., and Wooten, J.V., Measurement of methyl *tert*-butyl ether and tert-butyl alcohol in human blood and urine by purge-and-trap gas chromatography-mass-spectrometry using an isotope-dilution method, *Journal of Analytical Toxicology*, v. 19, no. 3, p. 187-191, 1995.

Borden, R.C., Intrinsic bioremediation of MTBE and BTEX—Field and laboratory results, in American Chemical Society Division of Environmental Chemistry pre-prints of papers, 213th, San Francisco, California, ACS, v. 37, no. 1, p. 426, 427, 1997.

Borden, R.C., Daniel, R.A., LeBrun, L.E., IV, and Davis, C.W., Intrinsic biodegradation of MTBE and BTEX in a gasoline-contaminated aquifer, *Water Resources Research*, v. 33, no. 5, p. 1105–1115, 1997.

Bott, D.J., Dawson, W.M., Piel, W.J., and Karas, L.J., MTBE environmental fate, London, England, Arco Chemical Company, The Institute of Petroleum, November 26, 1992.

Brady, J.F., Xiao, F., Ning, S.M., and Yang, C.S., Metabolism of methyl *tertiary*-butyl ether by rat hepatic microsomes: Archives of Toxicology, v. 64, no. 2, p. 157–160, 1990.

Bravo, H.A., Camacho, R.C., Roy-Ocotla, G.R., Sosa, R.E., and Torres, R.J., Analysis of the change in atmospheric urban formaldehyde and photochemistry activity as a result of using methyl-t-butyl-ether (MTBE) as an additive in gasolines of the metropolitan area of Mexico City, *Atmospheric Environment*, v. 25B, no. 2, p. 285–288, 1991.

Brown, A., Farrow, J.R.C., Rodriguez, R.A., Johnson, B.J., and Bellomo, A.J., Methyl *tertiary*-butyl ether (MtBE) contamination of the city of Santa Monica drinking water supply, in Stanley, A., Ed., NWWA/API Petroleum Hydrocarbons and Organic Chemicals in Groundwater—Prevention, Detection, and Remediation Conference, Houston, Tex., Nov. 12-14, Proceedings, Houston, Texas, National Water Well Association and American Petroleum Institute, p. 35–39, 1997.

Brown, A., Farrow, J.R.C., Rodriguez, R.A., and Johnson, B.J., Methyl *tertiary*-butyl ether (MTBE) contamination of the city of Santa Monica drinking water supply — an update, in The Southwest Focused Groundwater Conference—Discussing the Issue of MTBE and Perchlorate in Groundwater, Anaheim, Calif., June 3-4, Proceedings, Anaheim, California, National Groundwater Association, p. 16–25, 1998.

Bruce, B.W., Denver's urban groundwater quality—Nutrients, pesticides, and volatile organic compounds, U.S. Geological Survey Fact Sheet FS-106-95, 1995.

Bruce, B.W. and McMahon, P.B., Shallow groundwater quality beneath a major urban center—Denver, Colorado, USA, *Journal of Hydrology*, v. 186, p. 129–151, 1996.

Buchholtz, W.F. and Crow, W.L., Relating SARA Title III emissions to community exposure through ambient air quality measurements, in Air & Waste Management Association, Annual Meeting & Exhibition, 83rd, Pittsburgh, Pennsylvania, June 24-29, 1990, Proceedings, AWMA, p. 2–15, 1990.

Burbacher, T.M., Neurotoxic effects of gasoline and gasoline constituents, *Environmental Health Perspectives Supplements*, v. 101, suppl. 6, p. 133–141, 1993.

Burleigh-Flayer, H.D., Chun, J.S., and Kintigh, W.J., Methyl *tertiary*-butyl ether—Vapor inhalation oncogenicity study in CD-1 mice: Export, Pennsylvania, Bushy Run Research Center, BRRC report 91N0013A, 1992.

Buschek, T.E., Gallagher, D.J., Peargin, T.R., Kuehne, D.L., and Zuspan, C.R., Occurrence and behavior of MTBE in groundwater, in The Southwest Focused Groundwater Conference—Discussing the Issue of MTBE and Perchlorate in Groundwater, Anaheim, Calif., June 3-4, Proceedings, Anaheim, California, National Groundwater Association, p. 2–3, 1998.

Butillo, J.V., Pulido, A.D., Reese, N.M., and Lowe, M.A., Removal efficiency of MTBE in water—Confirmation of a predictive model through applied technology, in Stanley, A., Ed., NWWA/API Petroleum Hydrocarbons and Organic chemicals in Groundwater—Prevention, Detection, and Remediation Conference, Houston, Texas, November 2-4, Proceedings, Houston, Texas, National Water Well Association and American Petroleum Institute, p. 91–105, 1994.

Buxton, H.T., Landmeyer, J.E., Baer, A.L., Church, C.D., and Tratnyek, P.G., Interdisciplinary investigation of subsurface contaminant transport and fate at point-source releases of gasoline containing MTBE, in Stanley, A., Ed., NWWA/API Petroleum Hydrocarbon Conference—Prevention, Detection, and Restoration, Houston, Texas, Nov. 12-14, Proceedings, National Water Well Association and American Petroleum Institute, p. 2–18, 1997.

Cabrera, A.E., and Galindo, M.A., Preliminary evaluation of oxygenated fuels in a laboratory engine at Mexico City, in Air & Waste Management Association, Annual Meeting & Exhibition, 83rd, Pittsburgh, Pennsylvania, June 24-29, Proceedings, AWMA, 1990.

Cain, W.S., Leaderer, B.P., Ginsberg, G.L., Andrews, L.S., Cometto-Muniz, J.E., Gent, J.F., Buck, M., Berglund, L.G., Mohsenin, V., Monahan, E., and Kjaergaard, S., Acute exposure to low-level methyl *tertiary*-butyl ether (MTBE)—Human reactions and pharmacokinetic response, *Inhalation Toxicology*, v. 8, no. 1, p. 21–48, 1996.

Calvert, J.G., Heywood, J.B., Sawyer, R.F., and Seinfeld, J.H., Achieving acceptable air quality—Some reflections on controlling vehicle emissions, *Science*, v. 261, no. 5117, p. 37–45, 1993.

Canadian Environmental Protection Act, Priority substances list, assessment report no. 5, methyl *tertiary*-butyl ether: Government of Canada, Beauregard Printers Limited, 1992.

Carpenter, P.L. and Vinch, C.A., Remediation of overlapping benzene/MTBE and MTBE-only plumes—A case study, in Stanley, A., Ed., NWWA/API Petroleum Hydrocarbons and Organic Chemicals in Groundwater—Prevention, Detection, and Remediation Conference, Houston, Texas, National Water Well Association and American Petroleum Institute, p. 74–88, 1997.

Chatin, L., Fombarlet, C., Bernasconi, C., Gauthier, A., and Schmelzle, P., ETBE as a gasoline blending component—The experience of Elf Aquitaine: Society of Automotive Engineering, Spec. Publ. SP-1054 (Gasoline—Composition and additives to meet the performance and emission requirements of the nineties), p. 1–10, 1994.

Growth continues in chemical production, *Chemical & Engineering News*, v. 72, no. 27, p. 30–36, 1994.

RFG opening up pitfalls for oxygenates producers, *Chemical Marketing Reporter*, v. 242, no. 16, p. 7, 19, 1992.

Chen, C.S. and Delfino, J.J., Facilitated solubilization of polynuclear aromatic hydrocarbons by the co-solvent effect of oxygenated fuel additives and alternative fuels, abs., in American Chemical Society Division of Environmental Chemistry pre-prints of extended abstracts, 212th, Orlando, Fl., ACS, v. 36, no. 2, p. 312–314, 1996.

Chiang, C.Y., Loos, K.R., and Klopp, R.A., Field determination of geological/chemical properties of an aquifer by cone penetrometry and headspace analysis, *Groundwater*, v. 30, no. 3, p. 428–436, 1992.

Church, C.D., Isabelle, L.M., Pankow, J.F., Tratnyek, P.G., and Rose, D.L., Assessing the in situ degradation of methyl *tert*-butyl ether (MTBE) by product identification at the sub-ppb level using direct aqueous injection GC/MS, in American Chemical Society Division of Environmental Chemistry pre-prints of papers, 213th, San Francisco, California, ACS, v. 37, no. 1, p. 411–413, 1997.

Church, C.D., Isabelle, L.M., Pankow, J.F., Rose, D.L., and Tratnyek, P.G., Method for determination of methyl *tert*-butyl ether (MTBE) and its degradation products in water, *Environmental Science & Technology*, v. 31, no. 12, p. 3723–3726, 1997.

Clark, C.R., Dutcher, J.S., Henderson, T.R., McClellan, R.O., Marshall, W.F., Naman, T.M., and Seizinger, D.E., Mutagenicity of automotive particulate exhaust—Influence of fuel extenders, additives, and aromatic content, in MacFarland, H.N., Holdsworth, C.E., MacGregor, J.A., Call, R.W., and Lane, M.L., Eds., *Applied Toxicology of Petroleum Hydrocarbons*, Princeton, N.J., Princeton Scientific Publishers, p. 109–122, 1984.

Clark, J.J., Rodriguez, R.A., Brown, A., and Johnson, B.J., Public health implications of MTBE and perchlorate in water—Risk management decisions for water purveyors, in The Southwest Focused Groundwater Conference—Discussing the Issue of MTBE and Perchlorate in Groundwater, Anaheim, California, National Groundwater Association, p. 67–70, 1998.

Clayton Environmental Consultants, Gasoline vapor exposure assessment at service stations, Washington, D.C., Health and Environmental Sciences Department, American Petroleum Institute Publication No. 4553, May, 1993.

Cline, P.V., Delfino, J.J., and Rao, P.S.C., Partitioning of aromatic constituents into water from gasoline and other complex solvent mixtures, *Environmental Science & Technology*, v. 25, no. 5, p. 914–920, 1991.

Cochrane, R.A. and Hillman, D.E., Direct gas chromatographic determination of alcohols and methyl *tert*-butyl ether in gasolines using infrared detection, *Journal of Chromatography*, v. 287, no. 1, p. 197–201, 1984.

Cohen, Y., Partitioning of organic pollutants in the environment, in Managing Hazardous Air Pollutants—First International Conference, Washington, D.C., November 4-6, 1991, Chelsea, Michigan, Lewis Publishers, Inc., p. 278–295, 1993.

Coker, D.T., van den Hoed, N., Saunders, K.J., and Tindle, P.E., A monitoring method for gasoline vapor giving detailed composition, *Annual Occupational Hygiene*, v. 33, no. 1, p. 15–26, 1989.

Colucci, J.M. and Benson, J.D., 1992 [1993], Impact of reformulated gasoline on emissions from current and future vehicles, in Strauss, K.H. and Dukek, W.G., Eds., The Impact of U.S. Environmental Regulations on Fuel Quality, Ann Arbor, Mich., American Society for Testing and Materials, p. 105–123.

Conaway, C.C., Schroeder, R.E., and Snyder, N.K., Teratology evaluation of methyl *tertiary*-butyl ether in rats and mice, *Journal of Toxicology and Environmental Health*, v. 16, no. 6, p. 797–809, 1985.

Connor, B.F., Rose, D.L., Noriega, M.C., Murtagh, L.K., and Abney, S.R., Methods of analysis by the U.S. Geological Survey National Water Quality Laboratory, including detections less than reporting limits, U.S. Geological Survey Open-File Report OFR 97-829, 78 p., 1997.

Cook, J.R., Enns, P., and Sklar, M.S., Impact of the oxyfuel program on ambient CO levels, in American Chemical Society Division of Environmental Chemistry pre-prints of papers, 213th, San Francisco, California, ACS, v. 37, no. 1, p. 379–381, 1997.

Cornitius, T., California air rules foster MTBE demand, *Chemical Week*, v. 158, no. 27, p. 33, 1996.

Cox, R.A. and Goldstone, A., Atmospheric reactivity of oxygenated motor fuel additives, in Versino, B. and Ott, H., Eds., Physico-chemical behavior of atmospheric pollutants, European Symposium, 2nd, Varese, Italy, September 29-October 1, Proceedings: Boston, Mass., D. Reidel Publishing Company, p. 112–119, 1981.

Crowley, J.S. and Tulloch, C., Santa Clara Valley Water District's leaking UST oversight program, MTBE issues in Santa Clara County groundwater supplies, in The Southwest Focused Groundwater Conference—Discussing the Issue of MTBE and Perchlorate in Groundwater, Anaheim, California, June 3-4, 1998, Proceedings: Anaheim, California, National Groundwater Association, p. 26–35, 1998.

Dale, M.S., Losee, R.F., Crofts, E.W., and Davis, M.K., MTBE—Occurrence and fate in source-water supplies, in American Chemical Society Division of Environmental Chemistry pre-prints of papers, 213th, San Francisco, California, ACS, v. 37, no. 1, p. 276–377, 1997.

Daly, M.H. and Lindsey, B.D., Occurrence and concentrations of volatile organic compounds in shallow groundwater in the Lower Susquehanna River Basin, Pennsylvania and Maryland, Denver, CO, U.S. Geological Survey Water-Resources Investigations Report 96-4141, June, 1986.

Daniel, R.A., Intrinsic bioremediation of BTEX and MTBE—Field, laboratory and computer modeling studies, Raleigh, North Carolina State University, Master's thesis, 1995.

Davidson, J.M., Fate and transport of MTBE—The latest data, in Stanley, A., Ed., NWWA/API Petroleum Hydrocarbons & Organic Chemicals in Groundwater—Prevention, Detection, and Remediation Conference, Houston, Tex., Proceedings, Houston, Texas, National Water Well Association and American Petroleum Institute, p. 285–301, 1995a.

Davidson, J.M., Groundwater health issues of MTBE—Sources, MTBE in precipitation, MTBE in groundwater, fate & transport, MTBE in drinking water, Reformulated Gasoline Workshop, Proceedings, October 12, 1995b.

Davidson, J.M. and Parsons, R., Remediating MTBE with current and emerging technologies, in Stanley, A., Ed., NWWA/API Petroleum Hydrocarbons and Organic Chemicals in Groundwater—Prevention, Detection, and Remediation Conference, Houston, Tex., Nov. 13–15, Proceedings, Houston, Texas, National Water Well Association and American Petroleum Institute, p. 15–29, 1996.

Delzer, G.C., Zogorski, J.S., Lopes, T.J., and Bosshart, R.L., Occurrence of the gasoline oxygenate MTBE and BTEX compounds in urban stormwater in the United States, 1991-95, U.S. Geological Survey Water-Resources Investigation Report WRIR 96-4145, 1996.

Delzer, G.C., Zogorski, J.S., and Lopes, T.J., Occurrence of the gasoline oxygenate MTBE and BTEX compounds in municipal stormwater in the United States, 1991-95, in American Chemical Society Division of Environmental Chemistry pre-prints of papers, 213th, San Francisco, California, ACS, v. 37, no. 1, p. 374–376, 1997.

Denton, J. and Mazur, L., California's cleaner burning gasoline and methyl *tertiary*-butyl ether (abs.), in Society of Environmental Toxicology and Chemistry abstract book, 17th, Washington, D.C., November 17-21, Washington, D.C., SETAC, p. 115, 1996.

Department of the Environment, and Department of National Health and Welfare, Assessment of the priority substance methyl *tertiary*-butyl ether—Extract, Canada Gazette, Part I, January 30, p. 262–264, 1993.

Diehl, J.W., Finkbeiner, J.W., and DiSanzo, F.P., Determination of ethers and alcohols in reformulated gasolines by gas chromatography/atomic emission detection, *Journal of High Resolution Chromatography*, v. 18, no. 2, p. 108–110, 1995.

Drew, R.T., Misunderstood MTBE, *Environmental Health Perspectives*, v. 103, no. 5, p. 420, 1995.

Drobat, P.A., Bleckrnann, C.A., and Agrawal, A., Determination of the cometabolic biodegradation potential of methyl *tertiary*-butyl ether in laboratory microcosms, in American Chemical Society Division of Environmental Chemistry pre-prints of papers, 213th, San Francisco, California, ACS, v. 37, no. 1, p. 405, 406, 1997.

Duffy, J.S., Del Pup, J.A., and Kneiss, J.J., Toxicological evaluation of methyl *tertiary*-butyl ether (MTBE)—Testing performed under TSCA consent agreement, *Journal of Soil Contamination*, v. 1, no. 1, p. 29–37, 1992.

Dvorak, B.I., Lawler, D.F., Speitel, G.E., Jr., Jones, D.L., and Broadway, D.A., Selecting among physical/chemical processes for removing synthetic organics from water, *Water Environment Research*, v. 65, no. 7, p. 827–838, 1993.

Eweis, J.B., Watanabe, N., Schroeder, E.D., Chang, D.P.Y., and Scow, K.M., MTBE biodegradation in the presence of other gasoline components, in The Southwest Focused Groundwater Conference—Discussing the Issue of MTBE and Perchlorate in Groundwater, Anaheim, Calif., June 3-4, Proceedings, Anaheim, California, National Groundwater Association, p. 55–62, 1998.

Fiedler, N., Mohr, S.N., Kelly-McNeil, K., and Kipen, H.M., Response of sensitive groups to MTBE, *Inhalation Toxicology*, v. 6, no. 6, p. 539–552, 1994.

Freed, C.N., EPA fuel programs, in American Chemical Society Division of Environmental Chemistry pre-prints of papers, 213th, San Francisco, California, ACS, v. 37, no. 1, p. 366–368, 1997.

Fuels for the Future, A report on clean renewable fuels vs. dirty contaminated water and Liberty Village, Sullivan County, New York, Washington, D.C., November, 1994.

Fujiwara, Yasuo, Konoshita, T., Sato, H., and Kojima, I., Biodegradation and bioconcentration of alkyl ethers, *Yukagatu*, v. 33, no. 2, p. 111–115, Abs. and illus. in English, 1984.

Garrett, P., Moreau, M., and Lowry, J.D., MTBE as a groundwater contaminant, in NWWA/API Conference on Petroleum Hydrocarbons and Organic Chemicals in Groundwater—Prevention, Detection, and Restoration, Houston, Texas, November 12-14, Proceedings, Dublin, Ohio, National Water Well Association, p. 227–238, 1986.

Gerry, F.S., Schubert, A.J., McNally, M.J., and Pahl, R.H., Test fuels—Formulation and analyses—The auto/oil air quality improvement research program, Society of Automotive Engineers paper 920324, p. 335–357, 1992.

Gilbert, C.E. and Calabrese, E.J., Developing a standard for methyl *tertiary*-butyl ether in drinking water, in Gilbert, C.E. and Calabrese, E.J., Eds., *Regulating Drinking Water Quality*, Ann Arbor, Mich., Lewis Publishers, Inc., p. 231–252, 1992.

Gomez-Taylor, M.M., Abernathy, C.O., and Du, J.T., Drinking water health advisory for methyl *tertiary*-butyl ether, in American Chemical Society Division of Environmental Chemistry pre-prints of papers, 213th, San Francisco, California, ACS, v. 37, no. 1, p. 370–372, 1997.

Grady, S.J., Detections of MTBE in surficial and bedrock aquifers in New England (abs.), in Society of Environmental Toxicology and Chemistry Annual Meeting abstract book, 17th, Washington, D.C., November 17-21, Washington, D.C., SETAC, p. 115, 1996.

Grady, S.J., Distribution of MTBE in groundwater in New England by aquifer type and land use, in American Chemical Society Division of Environmental Chemistry pre-prints of papers, 213th, San Francisco, California, ACS, v. 37, no. 1, p. 392-394, 1997.

Grady, S.J., Volatile organic compounds in groundwater in the Connecticut, Housatonic, and Thames River Basins, 1993-1995, U.S. Geological Survey Fact Sheet FS-029-97, 1997.

Grady, S.J. and Mullaney, J.R., Natural and human factors affecting shallow water quality in surficial aquifers in the Connecticut, Housatonic, and Thames River Basins, U.S. Geological Survey Water-Resources Investigations Report, WRIR 98-4042, 1998.

Graves, K.L. and MacLeod, N.S., A basin protection strategy for sites with MTBE impacts, in The Southwest Focused Groundwater Conference—Discussing the Issue of MTBE and Perchlorate in Groundwater, Anaheim, Calif., June 3-4, Proceedings, Anaheim, California, National Groundwater Association, p. 71–75, 1998.

Green, A., Paillet, F.L., and Gurrieri, J.T., A multi-faceted evaluation of a gasoline contaminated bedrock aquifer in Connecticut (abs.) Geological Society of America Abstracts with Programs, v. 24, no. 3, p. 25, 1992.

Gregorski, D., Special study report—Investigating the relationship between the use of a gasoline additive (MTBE) and ambient air formaldehyde levels, Hartford, Conn., Bureau of Air Management, Monitoring and Radiation Division, Department of Environmental Protection, 1995.

Grosjean, D., Grosjean, E., and Rasmussen, R.A., Atmospheric chemistry and urban air concentrations of MTBE and ethanol, in American Chemical Society Division of Environmental Chemistry pre-prints of papers, 213th, San Francisco, California, ACS, v. 37, no. 1, p. 378–379, 1997.

Grosjean, E., Grosjean, D., Gunawardena, R., and Rasmussen, R.A., Ambient concentrations of ethanol and methyl *tert*-butyl ether in Porto Alegre, Brazil, March 1996-April 1997, *Environmental Science & Technology*, v. 32, no. 6, p. 736–742, 1998.

Groves, F.R., Jr., Effect of co-solvents on the solubility of hydrocarbons in water, *Environmental Science & Technology*, v. 22, no. 3, p. 282–286, 1988.

Halden, R.U., Schoen, S.R., Galperin, Y., Kaplan, I.R., and Happel, A.M., Evaluation of EPA and ASTM methods for analysis of oxygenates in gasoline—Contaminated groundwater, in The Southwest Focused Groundwater Conference—Discussing the Issue of MTBE of Perchlorate in Groundwater, Anaheim, Calif., June 3-4, Proceedings, Anaheim, California, National Groundwater Association, p. 1, 1998.

Hall, J.R., Part 2—Cleaner products—A refining challenge, *Hydrocarbon Processing*, v. 71, no. 5, p. 100-C–100-F, 1992.

Happel, A.M., Beckenbach, E., Savalin, L., Temko, H., Rempel, R., Dooher, B., and Rice, D., Analysis of dissolved benzene plumes and methyl *tertiary*-butyl ether (MTBE) plumes in groundwater at leaking underground fuel tank (LUFT) sites, in American Chemical Society Division of Environmental Chemistry pre-prints of papers, 213th, San Francisco, California, ACS, v. 37, no. 1, p. 409–411, 1997.

Happel, A.M., Beckenbach, E.H., and Halden, R.U., An evaluation of MTBE impacts to California groundwater resources, Livermore, California, Lawrence Livermore National Laboratory, UCRL-AR-130897, 1998.

Hardisty, P.E. and Schroder, R.A., Fracture-controlled transport of MTBE, in Stanley, A., Ed., NWWA/API Petroleum Hydrocarbons and Organic Chemicals in Groundwater—Prevention, Detection, and Remediation Conference, Houston, Texas, November 13-15, Proceedings, Houston, Texas, National Water Well Association and American Petroleum Institute, p. 31–44, 1996.

Hartle, R., Exposure to methyl *tert*-butyl ether and benzene among service station attendants and operators, *Environmental Health Perspectives Supplements*, v. 101, suppl. 6, p. 23–26, 1993.

Hartley, W.R. and Englande, A.J., Jr., Health risk assessment of the migration of unleaded gasoline—A model for petroleum products, *Water Science & Technology*, v. 25, no. 3, p. 65–72, 1992.

Hoekman, S.K., Speciated measurements and calculated reactivities of vehicle exhaust emissions from conventional and reformulated gasolines, *Environmental Science & Technology*, v. 26, no. 6, p. 1206-1216, 1992.

Hoekman, S.K., Improved gas chromatography procedure for speciated hydrocarbon measurements of vehicle emissions, *Journal of Chromatography*, v. 639, no. 2, p. 239–253, 1993.

Horan, C.M. and Brown, E.J., Biodegradation and inhibitory effects of methyl *tertiary*-butyl ether (MTBE) added to microbial consortia, in Annual Conference on Hazardous Waste Research, 10th, Manhattan, Kansas State University, May 23-24, Proceedings, p. 11–19, 1995.

Howard, P.H., Boethling, R.S., Jarvis, W.F., Meylan, W.M., and Michalenko, E.M., *Handbook of Environmental Degradation Rates*, Chelsea, Michigan, Lewis Publishers, Inc., p. 653–654, 1991.

Howard, P.H., Ed., *Handbook of Environmental Fate and Exposure Data for Organic Chemicals*, Ann Arbor, Mich., Lewis Publishers, Inc., v. IV, p. 71–75, 1993.

Hsieh, C.R. and Ouimette, J.R., 1994, Comparative study of multimedia modeling for dynamic partitioning of fossil fuels-related pollutants, *Journal of Hazardous Materials*, v. 37, no. 3, p. 489–505.

Hubbard, C.E., Barker, J.F., and Vandegriendt, M., Transport and fate of dissolved methanol, methyl *tertiary*-butyl-ether, and monoaromatic hydrocarbons in a shallow sand aquifer—Appendix H—Laboratory biotransformation studies, Washington, D.C., Health and Environmental Sciences Department, American Petroleum Institute Publication No. 4601, 1994.

Hubbard, C.E., Barker, J.F., O'Hannesin, S.F., Vandegriendt, M., and Gillham, R.W., Transport and fate of dissolved methanol, methyl *tertiary*-butyl-ether, and monoaromatic hydrocarbons in a shallow sand aquifer, Washington, D.C., Health and Environmental Sciences Department, American Petroleum Institute Publication No. 4601, 1994.

Hunt, C.S., Cronkhite, L.A., Corseuil, H.X., and Alvarez, P.J.J., Effect of ethanol on anaerobic toluene degradation in aquifer microcosms, in American Chemical Society Division of Environmental Chemistry pre-prints of papers, 213th, San Francisco, California, ACS, v. 37, no. 1, p. 424–426, 1997.

Hutcheon, D.E., Arnold, J.D., ten Hove, W., and Boyle, J., III, Disposition, metabolism, and toxicity of methyl tertiary butyl ether, an oxygenate for reformulated gasoline, *Journal of Toxicology and Environmental Health*, v. 47, no. 5, p. 453–464, 1996.

Iborra, M., Izquierdo, J.F., Tejero, J., and Cunill, F., Getting the lead out with ethyl t-butyl ether, *Chemtech*, v. 18, no. 2, p. 120–122, 1988.

International Technology Corporation, Treatment system for the reduction of aromatic hydrocarbons and ether concentrations in groundwater, Washington, D.C., Health and Environmental Sciences Department, American Petroleum Institute Publication No. 4471, 1988.

Jandrasi, F.J. and Masoomian, S.Z., Minimize process waste during plant design, *Environmental Engineering World*, v. 1, no. 1, p. 6–15, 1995.

Japar, S.M., Wallington, T.J., Richert, J.F.O., and Ball, J.C., The atmospheric chemistry of oxygenated fuel additives—t-Butyl alcohol and t-butyl ether, in Air & Waste Management Association, Annual Meeting & Exhibition, 83rd, Pittsburgh, Pennsylvania, June 24-29, Proceedings, AWMA, v. 6, 1990a.

Japar, S.M., Wallington, T.J., Richert, J.F.O., and Ball, J.C., The atmospheric chemistry of oxygenated fuel additives—t-Butyl alcohol, dimethyl ether, and methyl t-butyl ether, *International Journal of Chemical Kinetics*, v. 22, no. 12, p. 1257–1269, 1990b.

Japar, S.M., Wallington, T.J., Rudy, S.J., and Chang, T.Y., Ozone-forming potential of a series of oxygenated organic compounds, *Environmental Science & Technology*, v. 25, no. 3, p. 415–420, 1991.

Javanmardian, M. and Glasser, H.A., In-situ biodegradation of MTBE using biosparging, in American Chemical Society Division of Environmental Chemistry pre-prints of papers, 213th, San Francisco, California, ACS, v. 37, no. 1, p. 424, 1997.

Horan, C.M. and Brown, E.J., Biodegradation and inhibitory effects of methyl *tertiary*-butyl ether (MTBE) added to microbial consortia, in Annual Conference on Hazardous Waste Research, 10th, Manhattan, Kansas State University, May 23-24, Proceedings, p. 11–19, 1995.

Howard, P.H., Boethling, R.S., Jarvis, W.F., Meylan, W.M., and Michalenko, E.M., *Handbook of Environmental Degradation Rates*, Chelsea, Michigan, Lewis Publishers, Inc., p. 653-654, 1991.

Howard, P.H., Ed., *Handbook of Environmental Fate and Exposure Data for Organic Chemicals*, Ann Arbor, Mich., Lewis Publishers, Inc., v. IV, p. 71-75, 1993.

Hsieh, C.R. and Ouimette, J.R., 1994, Comparative study of multimedia modeling for dynamic partitioning of fossil fuels-related pollutants, *Journal of Hazardous Materials*, v. 37, no. 3, p. 489–505.

Hubbard, C.E., Barker, J.F., and Vandegriendt, M., Transport and fate of dissolved methanol, methyl *tertiary*-butyl-ether, and monoaromatic hydrocarbons in a shallow sand aquifer—Appendix H—Laboratory biotransformation studies, Washington, D.C., Health and Environmental Sciences Department, American Petroleum Institute Publication No. 4601, 1994.

Hubbard, C.E., Barker, J.F., O'Hannesin, S.F., Vandegriendt, M., and Gillham, R.W., Transport and fate of dissolved methanol, methyl *tertiary*-butyl-ether, and monoaromatic hydrocarbons in a shallow sand aquifer, Washington, D.C., Health and Environmental Sciences Department, American Petroleum Institute Publication No. 4601, 1994.

Hunt, C.S., Cronkhite, L.A., Corseuil, H.X., and Alvarez, P.J.J., Effect of ethanol on anaerobic toluene degradation in aquifer microcosms, in American Chemical Society Division of Environmental Chemistry pre-prints of papers, 213th, San Francisco, California, ACS, v. 37, no. 1, p. 424–426, 1997.

Hutcheon, D.E., Arnold, J.D., ten Hove, W., and Boyle, J., III, Disposition, metabolism, and toxicity of methyl tertiary butyl ether, an oxygenate for reformulated gasoline, *Journal of Toxicology and Environmental Health*, v. 47, no. 5, p. 453–464, 1996.

Iborra, M., Izquierdo, J.F., Tejero, J., and Cunill, F., Getting the lead out with ethyl t-butyl ether, *Chemtech*, v. 18, no. 2, p. 120–122, 1988.

International Technology Corporation, Treatment system for the reduction of aromatic hydrocarbons and ether concentrations in groundwater, Washington, D.C., Health and Environmental Sciences Department, American Petroleum Institute Publication No. 4471, 1988.

Jandrasi, F.J. and Masoomian, S.Z., Minimize process waste during plant design, *Environmental Engineering World*, v. 1, no. 1, p. 6–15, 1995.

Japar, S.M., Wallington, T.J., Richert, J.F.O., and Ball, J.C., The atmospheric chemistry of oxygenated fuel additives—t-Butyl alcohol and t-butyl ether, in Air & Waste Management Association, Annual Meeting & Exhibition, 83rd, Pittsburgh, Pennsylvania, June 24-29, Proceedings, AWMA, v. 6, 1990a.

Japar, S.M., Wallington, T.J., Richert, J.F.O., and Ball, J.C., The atmospheric chemistry of oxygenated fuel additives—t-Butyl alcohol, dimethyl ether, and methyl t-butyl ether, *International Journal of Chemical Kinetics*, v. 22, no. 12, p. 1257–1269, 1990b.

Japar, S.M., Wallington, T.J., Rudy, S.J., and Chang, T.Y., Ozone-forming potential of a series of oxygenated organic compounds, *Environmental Science & Technology*, v. 25, no. 3, p. 415–420, 1991.

Javanmardian, M. and Glasser, H.A., In-situ biodegradation of MTBE using biosparging, in American Chemical Society Division of Environmental Chemistry pre-prints of papers, 213th, San Francisco, California, ACS, v. 37, no. 1, p. 424, 1997.

Jeffrey, D., Physico-chemical properties of MTBE and predictions of preferred environmental fate and compartmentalization, in American Chemical Society Division of Environmental Chemistry pre-prints of papers, 213th, San Francisco, California, ACS, v. 37, no. 1, p. 397–399, 1997.

Jensen, H.M. and Arvin, E., Solubility and degradability of the gasoline additive MTBE, methyl-*tert*-butyl-ether, and gasoline compounds in water, in Arendt, F., Hinsenveld, M., and van den Brink, W.J., Eds., *Contaminated Soil '90*, Netherlands, Kluwer Academic Publishers, p. 445–448, 1990.

Johansen, N.G., The analysis of C1–C4 alcohols, MTBE, and DIPE in motor gasolines by multi-dimensional capillary column gas chromatography, *Journal of High Resolution Chromatography & Chromatography Communications*, v. 7, no. 8, p. 487–489, 1984.

Johnson, T., McCoy, M., and Wisbith, T., A study to characterize air concentrations of methyl *tertiary*-butyl ether (MTBE) at service stations in the Northeast, Washington, D.C., Health and Environmental Sciences Department, American Petroleum Institute Publication No. 4619, 1994.

Johnson, R.L. and Grady, D.E., Remediation of a fractured clay soil contaminated with gasoline containing MTBE, in Stanley, A., Ed., NWWA/API Petroleum Hydrocarbons and Organic Chemicals in Groundwater—Prevention, Detection, and Remediation Conference, Houston, Texas, Nov. 12-14, Proceedings, Houston, Texas, National Water Well Association and American Petroleum Institute, p. 60–73, 1997.

Kanai, H., Inouye, V., Goo, R., Chow, R., Yazawa, L., and Maka, J., GC/MS analysis of MTBE, ETBE, and TAME in gasolines, *Analytical Chemistry*, v. 66, no. 6, p. 924–927, 1994.

Karpel Vel Leitner, N., Papailhou, A.L., Croue, J.P., Peyrot, J., and Dore, M., Oxidation of methyl *tert*-butyl ether (MTBE) and ethyl tert-butyl ether (ETBE) by ozone and combined ozone/hydrogen peroxide, *Ozone Science & Engineering*, v. 16, no. 1, p. 41–54, 1994.

Kelly, T.J., Callahan, P.J., Plell, J., and Evans, G.F., Method development and field measurements for polar volatile organic compounds in ambient air, *Environmental Science & Technology*, v. 27, no. 6, p. 1146–1153, 1993.

Kemezis, P., Precursors for MTBE/TAME, *Chemical Week*, v. 151, no. 1, p. 48, 1992.

Kirchstetter, T.W., Singer, B.C., Harley, R.A., Kendall, G.R., and Chan, W., Impact of oxygenated gasoline use on California light-duty vehicle emissions, *Environmental Science & Technology*, v. 30, no. 2, p. 661–670, 1996.

Kirschner, E., Alaska, Boston plan no-MTBE winter, *Chemical Week*, v. 153, no. 12, p. 7, 1993.

Kirschner, E.M., Production of top 50 chemicals increased substantially in 1994, *Chemical & Engineering News*, v. 73, no. 15, p. 16–18, 20, 1995.

Klan, M.J. and Carpenter, M.J., A risk-based drinking water concentration for methyl *tertiary*-butyl ether (MTBE), in Stanley, A., Ed., NWWA/API Petroleum Hydrocarbons and Organic Chemicals in Groundwater—Prevention, Detection, and Remediation Conference, Houston, Texas, November 2-4, Houston, Texas, National Water Well Association and American Petroleum Institute, p. 107-115, 1994.

Kolpin, D.W., Squillance, P.J., Zogorski, J.S., and Barbash, J.E., Pesticides and volatile organic compounds in shallow urban groundwater of the United States, in Chilton, J. et al., Eds., Congress on Groundwater in the Urban Environment Proceedings, Netherlands, A.A., Balkema, p. 469–474, 1997.

Komex H2O Science, Draft investigation report—MTBE contamination—City of Santa Monica Charnock well field, Los Angeles, California; Huntington Beach, California, 1997.

Lacy, M.J., Robbins, G.A., Wang, S., and Stuart, J.D., Use of sequential purging with the static headspace method to quantify gasoline contamination, *Journal of Hazardous Materials*, v. 43, no. 1-2, p. 31–44, 1995.

Landmeyer, J.E., Chapelle, F.H., and Bradley, P.M., Assessment of intrinsic bioremediation of gasoline contamination in the shallow aquifer, Laurel Bay Exchange, Marine Corps Air Station, Beaufort, South Carolina, U.S. Geological Survey Water-Resources Investigations Report 96-4026, 1996.

Landmeyer, J.E., Pankow, J.F., and Church, C.D., Occurrence of MTBE and tert-butyl alcohol in a gasoline-contaminated aquifer, in American Chemical Society Division of Environmental Chemistry pre-prints of papers, 213th, San Francisco, California, ACS, v. 37, no. 1, p. 413–415, 1997.

Landmeyer, J.E., Chappelle, F.H., Bradley, P.M., Pankow, J.F., Church, C.D., and Tratnyek, P.G., Fate of MTBE relative to benzene in a gasoline-contaminated aquifer (1993-98), *GroundWater Monitoring & Remediation*, Fall, 1998.

Lawuyi, R., and Fingas, M., Environmental impact of methyl *tert*-butyl ether (MTBE), in Technical Seminar of Chemical Spills, 14th, Proceedings, Ottawa, Ontario, Environment Canada, p. 127–141, 1997.

Lee, A.K.K. and Al-Jarallah, A., MTBE production technologies and economics, *Chemical Economy & Engineering Review*, v. 18, no. 9, p. 25–34, 1986.

Levy, J.M. and Yancey, J.A., Dual capillary gas chromatographic analysis of alcohols and methyl *tert*-butyl ether in gasolines, *Journal of High Resolution Chromatography & Chromatography Communications*, v. 9, no. 7, p. 383–387, 1986.

Lindsey, B.D., Breen, K.J., and Daly, M.H., MTBE in water from fractured-bedrock aquifers, south central Pennsylvania, in American Chemical Society Division of Environmental Chemistry pre-prints of papers, 213th, San Francisco, California, ACS, v. 37, no. 1, p. 399–400., 1997.

Lindstrom, A.B. and Pleil, J.D., Alveolar breath sampling and analysis to exposures to methyl *tertiary*-butyl ether (MTBE) during motor vehicle refueling, *Journal of Air & Waste Management*, v. 46, no. 7, p. 676–682, 1996.

Lioy, P.J., Weisel, C.P., Jo, W.K., Pellizzari, E., and Raymer, J.H., Microenvironmental and personal measurements of methyl *tertiary*-butyl ether (MTBE) associated with automobile use activities, *Journal of Exposure Analysis and Environmental Epidemiology*, v. 4, no. 4, p. 427–441, 1994.

Long, G., Meek, M.E., and Savard, S., Methyl *tertiary*-butyl ether—Evaluation of risks to health from environmental exposure in Canada, *Journal of Environmental Science and Health*, v. C12, no. 2, p. 389–395, 1994.

Lopes, T.J. and Bender, D.A., Nonpoint sources of volatile organic compounds in urban areas, in Source Water Assessment and Protection—A Technical Conference, Dallas, Texas, April 28-30 Proceedings, Fountain Valley, California National Water Institute, p. 199–200, 1998.

Lucas, A., Health concerns fuel EPA study of ETBE and TAME, *Chemical Week*, v. 154, no. 18, p. 10, 1994.

Lucier, G., Genter, M.B., Lao, Y.J., Stopford, W., and Starr, T., Summary of the carcinogenicity assessment of MTBE conducted by the Secretary's Scientific Advisory Board on Toxic Air Pollutants, *Environmental Health Perspectives*, v. 103, no. 5, p. 420–422, 1995.

Luhrs, R.C. and Pyott, C.J., Trilinear plots a powerful new application for mapping gasoline contamination, in NWWA/API Petroleum Hydrocarbons and Organic Chemicals in Groundwater—Prevention, Detection, and Restoration, Proceedings, Dublin, Ohio, National Water Well Association, p. 85–100, 1992.

Lyons, C.E., Quantifying the emissions reduction effectiveness and costs of oxygenated gasoline, in Air & Waste Management Association, Annual Meeting & Exhibition, 86th, Denver, Colorado, June 13-18, Proceedings, AWMA, 1993.

Lyons, C.E. and Fox, R.J., Quantifying the air pollution emissions reduction effectiveness and costs of oxygenated fuels, in Engineering, Society of Automotive, Ed., *New Developments in Alternative Fuels and Gasolines for SI and CI Engines*, Spec. Pub., SAE, SP-958, p. 61–75, 1993.

Mackay, D., Shiu, W.Y., and Ma, K.C., Illustrated handbook of physical-chemical properties and environmental fate for organic chemicals—Volume III—Volatile organic chemicals, Ann Arbor, Michigan, Lewis Publishers, Inc., p. 756, 1993.

Majima, T., Ishii, T., and Arai, S., The IR photochemistry of organic compounds. II. The IR photochemistry of ethers—The decomposition patterns, *Bulletin of the Chemical Society of Japan*, v. 62, no. 6, p. 1701–1709, 1989.

Malley, J.P., Jr., Eliason, P.A., and Wagler, J.L., Point-of-entry treatment of petroleum contaminated water supplies, *Water Environment Research*, v. 65, no. 2, p. 119–128, 1993.

Mancini, E.R., Aquatic toxicity data for methyl *tertiary*-butyl ether (MTBE)—Current status, future research, in American Chemical Society Division of Environmental Chemistry pre-prints of papers, 213th, San Francisco, California, ACS, v. 37, no. 1, p. 427–429, 1997.

Mancini, E.R., Physiochemical and ecotoxicological properties of gasoline oxygenates (abs.), Bridging the Global Environment—Technology, Communication, and Education, Society of Environmental Toxicology and Chemistry Annual Meeting Abstract Book, 18th, San Francisco, California, Nov. 16-20, Washington, D.C., SETAC, p. 251, 1997.

Mancini, E.R., Stubblefield, W.A., and Tillquist, H., Important ecological risk assessment parameters for MTBE and other gasoline oxygenates, in The Southwest Focused Groundwater Conference—Discussing the Issue of MTBE and Perchlorate in Groundwater, Anaheim, Calif., June 3-4, Proceedings, Anaheim, California, National Groundwater Association, p. 55–62, 1998.

Mannino, D.M., Schreiber, J., Aldous, K., Ashley, D., Moolenaar, R., and Almaguer, D., Human exposure to volatile organic compounds—A comparison of organic vapor monitoring badge levels with blood levels, *International Archives of Occupational Environmental Health*, v. 67, no. 1, p. 59–64, 1995.

Mannino, D.M. and Etzel, R.A., Are oxygenated fuels effective? An evaluation of ambient carbon monoxide concentrations in 11 western states, 1986 to 1992, *Journal of the Air & Waste Management Association*, v. 46, no. 1, p. 20–24, 1996.

Marston, C.R., Improve etherification plant efficiency and safety, *Fuel Reformulation*, v. 4, no. 4, p. 42–46, 1994.

McCabe, L.J., Initial results from the auto/oil air quality improvement research program, in Strauss, K.H. and Dukek, W.G., Eds., *The Impact of U.S. Environmental Regulations on Fuel Quality*, Philadelphia, Pennsylvania, American Society for Testing and Materials, p. 63–83, 1992.

McKinnon, R.J. and Dyksen, J.E., Removing organics from groundwater through aeration plus GAC, *Journal of the American Water Works Association*, v. 76, no. 5, p. 42–47, 1984.

McMahon, P.B. and Bruce, B.W., Distribution of terminal electron-accepting processes in an aquifer having multiple contaminant sources, *Applied Geochemistry*, v. 12, no. 4, p. 507–516, 1997.

McNair, L., Russell, A., and Odman, M.T., Airshed calculation of the sensitivity of pollutant formation to organic compound classes and oxygenates associated with alternative fuels, *Journal of the Air Waste Management Association*, v. 42, no. 2, p. 174–178, 1992.

Mehlman, M.A., Dangerous properties of petroleum-refining products—Carcinogenicity of motor fuels (gasoline), *Teratogenesis, Carcinogenesis, and Mutagenesis*, v. 10, no. 5, p. 399–408, 1990.

Mehlman, M.A., Dangerous and cancer-causing properties of products and chemicals in the oil refining and petrochemical industry, Part XV. Health hazards and health risks from oxygenated automobile fuels (MTBE)—Lessons not heeded, *International Journal of Occupational Medicine and Toxicology*, v. 4, no. 2, p. 17, 1995.

Mehlman, M.A., Collegium Ramazzini position on oxygenated and reformulated gasoline, *International Journal of Occupational Medicine, Immunology, and Toxicology*, v. 5, no. 1, p. 1, 2, 1996.

Mennear, J.H., MTBE—Not carcinogenic, *Environmental Health Perspectives*, v. 103, no. 11, p. 985–986, 1995.

Mihelcic, J.R., Modeling the potential effect of additives on enhancing the solubility of aromatic solutes contained in gasoline, *GWMR*, v. 10, no. 3, p. 132–137, 1990.

Mo, K., Lora, C.O., Wanken, A., and Kulpa, C.F., Biodegradation of methyl-t-butyl ether by pure bacterial cultures (abs.), in Abstracts of the 95th general meeting of the American Society for Microbiology, American Society for Microbiology, v. 95, p. 408, 1995.

Mohr, S.N., Fiedler, N., Weisel, C., and Kelly-McNeil, K., Health effects of MTBE among New Jersey garage workers, *Inhalation Toxicology*, v. 6, no. 6, p. 553–562, 1994.

Moolenaar, R.L., Hefflin, B.J., Ashley, D.L., Middaugh, J.P., and Etzel, R.A., Methyl *tertiary*-butyl ether in human blood after exposure to oxygenated fuel in Fairbanks, Alaska, *Archives of Environmental Health*, v. 49, no. 5, p. 402–409, 1994.

Mormile, M.R., Liu, S., and Suflita, J.M., Anaerobic biodegradation of gasoline oxygenates—Extrapolation of information to multiple sites and redox conditions, *Environmental Science & Technology*, v. 28, no. 9, p. 1727–1732, 1994.

Mosteller, D.C., Reardon, K.F., Bourquin, A.W., Desilets, B., Dumont, D., Hines, R., and Kilkenny, S., Biotreatment of MTBE-contaminated groundwater, in American Chemical Society Division of Environmental Chemistry pre-prints of papers, 213th, San Francisco, California, ACS, v. 37, no. 1, p. 420, 421, 1997.

Nakamura, D.N., Is MTBE losing its popularity? *Hydrocarbon Processing*, v. 72, no. 9, p. 19, 1993.

National Technical Information Service, Methyl *tertiary*-butyl ether, in Government reports announcements & index, no. 15, NTIS, 1992.

NATLSCO, A Division of KRMS, Service station personnel exposures to oxygenated fuel components—1994, Washington, D.C., Health and Environmental Sciences Department, American Petroleum Institute Publication No. 4625, 1995.

Newman, A., MTBE detected in survey of urban groundwater, *Environmental Science & Technology*, v. 29, no. 7, p. 305A, 1995.

Department of Health, State of New York Fact Sheet—Village of Liberty water supply system, 1993.

Nihlen, A., Lof, A., and Johanson, G., Liquid/air partition coefficients of methyl and ethyl t-butyl ethers, t-amyl methyl ether, and t-butyl alcohol, *Journal of Exposure Analysis and Environmental Epidemiology*, v. 5, no. 4, p. 573–582, 1995.

Nocca, J. -L., Forestiere, A., and Cosyns, J., Diversity process strategies for reformulated gasoline, *Fuel Reformulation*, v. 4, no. 5, p. 18–22, 1994.

Novak, J.T., Yeh, C., Gullic, D., Eichenberger, J., and Benoit, R.E., The influence of microbial ecology on subsurface degradation of petroleum contaminants, Blacksburg, VA, Virginia Polytechnic Institute and State University, VPI-VWRRC-BULL 177, 1992.

O'Brien, A.K., Reiser, R.G., and Gylling, H., Spatial variability of volatile organic compounds in streams on Long Island, New York and in New Jersey, U.S. Geological Survey Fact Sheet FS-194-97, 1997.

Office of Science and Technology Policy, Interagency assessment of oxygenated fuels, Washington, D.C., OSTP, The Executive Office of the President, p. 258, 1995.

Oil & Gas Journal, Sabic sees big growth for MTBE, v. 92, no. 48, p. 30-31, 1994.

Oil & Gas Journal, USGS reports MTBE in groundwater, v. 93, no. 16, p. 21–22, 1995.

U.S. gasoline plagued by economic, technical uncertainty, *Oil & Gas Journal*, v. 94, no. 2, p. 29–33, 1996.

Oxygenated Fuels Association, MTBE in groundwater—Fact sheet for local health and water authorities, 1995.

Oxygenated Fuels Association, Gasoline reformulated with methyl *tertiary*-butyl ether (MTBE)—Public health issues and answers: Arlington, Virginia, 1996.

Oxygenated Fuels Association, Modeling the volatilization of methyl *tertiary*-butyl ether (MTBE) from surface impoundments, Arlington, Virginia, 1998.

Oxygenated Fuels Association, Taste and odor properties of methyl *tertiary*-butyl ether (MTBE) from surface impoundments, Arlington, Virginia, 1998.

Page, N.P., Gasoline leaking from underground storage tanks—Impact on drinking water quality, in Hemphill, D.D., Ed., *Trace Substances in Environmental Health XXII*, Columbia, University of Missouri, p. 233–245, 1989.

Palassis, J., Hartle, R.W., and Holtz, J.L., A method for determination of methyl *tert*-butyl ether in gasoline vapors and liquid gasoline samples, *Applied Occupational Environmental Hygiene*, v. 8, no. 11, p. 964–969, 1993.

Pankow, J.F., Rathburn, R.E., and Zogorski, J.S., Calculated volatilization rates of fuel oxygenate compounds and other gasoline-related compounds from rivers and streams—A comparison with other gasoline-related compounds, *Chemosphere*, v. 33, no. 5, p. 921–937, 1996.

Pankow, J.F., Thomson, N.R., and Johnson, R.L., Modeling the atmospheric inputs of MTBE to groundwater systems (abs.), in Society of Environmental Toxicology and Chemistry abstract book, 17th, Washington, D.C., November 17-21, SETAC, p. 115, 1996.

Pankow, J.F., Thomson, N.R., Johnson, R.L., Baehr, A.L., and Zogorski, J.S., The urban atmosphere as a non-point source for the transport of MTBE and other volatile organic compounds (VOCs) to shallow groundwater, in American Chemical Society Division of Environmental Chemistry pre-prints of papers, 213th, San Francisco, California, ACS, v. 37, no. 1, p. 385–387, 1997.

Pankow, J.F., Thomson, N.R., Johnson, R.L., Baehr, A.L., and Zogorski, J.S., The urban atmosphere as a non-point source for the transport of MTBE and other volatile organic compounds (VOCs) to shallow groundwater, *Environmental Science & Technology*, v. 31, no. 10, p. 2821–2828, 1997.

Park, K. and Cowan, R.M., Effects of oxygen and temperature on the biodegradation of MTBE, in American Chemical Society Division of Environmental Chemistry pre-prints of papers, 213th, San Francisco, California, ACS, v. 37, no. 1, p. 421–424, 1997.

Patterson, G.J., Potential claims for water purveyors impacted by MTBE or perchlorate, in The Southwest Focused Groundwater Conference—Discussing the Issue of MTBE and Perchlorate in Groundwater, Proceedings, Anaheim, California, June 3-4, Groundwater Association, p. 78–85, 1998.

Paulov, S., Action of the anti-detonation preparation tert.-butyl methyl ether on the model species Rana temporaria L., *Biologia*, v. 42, no. 2, p. 185–189 (Abs. in English), 1987.

Pauls, R.E., Determination of high octane components—Methyl t-butyl ether, benzene, toluene, and ethanol in gasoline by liquid chromatography, *Journal of Chromatographic Science*, v. 23, no. 10, p. 437–441, 1985.

Pavne, R.E., Novick, N.J., and Gallagher, M.N., Demonstrating intrinsic bioremediation of MTBE and BTEX in groundwater at a service station site, in American Chemical Society Division of Environmental Chemistry pre-prints of papers, 213th, San Francisco, California, ACS, v. 37, no. 1, p. 418–419, 1997.

Peaff, G., Court ruling spurs continued debate over gasoline oxygenates, *Chemical & Engineering News*, v. 72, no. 39, p. 8–13, 1994.

Pearson, G. and Oudijk, G., Investigation and remediation of petroleum product releases from residential storage tanks, *Groundwater Monitoring Review*, v. 13, no. 3, p. 124–128, 1993.

Piel, W.J., MTBE use and possible occurrences in water supplies, ARCO Chemical Company, May 10, 1995.

Poore, M., Chang, B., Niyati, F., and Madden, S., Sampling and analysis of methyl t-butyl ether in ambient air at selected locations in California, in American Chemical Society Division of Environmental Chemistry pre-prints of papers, 213th, San Francisco, California, ACS, v. 37, no. 1, p. 407, 1997.

Post, G., Methyl *tertiary*-butyl ether health-based maximum contaminant level support document, in New Jersey Drinking Water Quality Institute, Maximum contaminant level recommendations for hazardous contaminants in drinking water—Appendix A—Health-based maximum contaminant level support documents and addenda, Trenton Division of Science and Research, New Jersey Department of Environmental Protection, July, 1994.

Poulsen, M., Lemon, L., and Barker, J.F., Dissolution of monoaromatic hydrocarbons into groundwater from gasoline-oxygenate mixtures, *Environmental Science & Technology*, v. 26, no. 12, p. 2483–2489, 1992.

Prager, J.C., Ed., Methyl *tert*-butyl ether, *Dangerous Properties of Industrial Materials Report*, v. 12, no. 3, p. 381–394, 1992.

Prah, J.D., Goldstein, G.M., Devlin, R., Otto, D., Ashley, D., House, D., Cohen, K.L., and Gerrity, T., Sensory, symptomatic, inflammatory, and ocular responses to and the metabolism of methyl *tertiary*-butyl ether in a controlled human exposure experiment, *Inhalation Toxicology*, v. 6, no. 6, p. 521–538, 1994.

Price, J., Gas is greener, but smog safer, *Insight*, v. 11, no. 16, p. 27, 1995.

Quigley, C.J., Allen, D.T., and Corsi, R.L., Release of MTBE and other reformulated gasoline vapor constituents during vehicle refueling and storage tank loading, in American Chemical Society Division of Environmental Chemistry pre-prints of papers, 213th, San Francisco, California, ACS, v. 37, no. 1, p. 384–385, 1997.

Raese, J.W., Sandstrom, M.W., and Rose, D.L., U.S. Geological Survey laboratory method for MTBE and other fuel additives, U.S. Geological Survey Fact Sheet FS-219-95, 1995.

Rathbun, R.E., Transport, behavior, and fate of volatile organic compounds in streams, U.S. Geological Survey Professional Paper 1589, 1998.

Reichhardt, T., A new formula for fighting urban ozone, *Environmental Science & Technology*, v. 29, no. 1, p. 36A-41A, 1995.

Reisch, M.S., Top 50 chemicals production rose modestly last year, *Chemical & Engineering News*, v. 72, no. 15, p. 12–16, 1994.

Rixey, W.G., The effect of oxygenated fuels on the mobility of gasoline components in groundwater, in Stanley, A., Ed., NWWA/API Petroleum Hydrocarbons and Organic Chemicalsin Groundwater—Prevention, Detection, and Remediation, Proceedings, Houston, Texas, November 2-4, National Water Well Association and American Petroleum Institute, p. 75–90, 1994.

Robbins, G.A., Wang, S., and Stuart, J.D., Using the static headspace method to determine Henry's Law constants, *Analytical Chemistry*, v. 65, no. 21, p. 3113–3118, 1993.

Robinson, M., Bruner, R.H., and Olson, G.R., Fourteen- and ninety-day oral toxicity studies of methyl *tertiary*-butyl ether in Sprague-Dawley rats, *Journal of the American College of Toxicology*, v. 9, no. 5, p. 525-540, 1990.

Rodriguez, R., MTBE—Clean water vs. clean air, Ontario, Canada, Association of California Water Agencies, March 13, 1997.

Rosenkranz, H.S. and Klopman, G., Predictions of the lack of genotoxicity and carcinogenicity in rodents of two gasoline additives—Methyl- and ethyl-t-butyl ethers, *In Vitro Toxicology*, v. 4, no. 1, p. 49, 50, 1991.

Rotman, D., Effects of oxygenated fuels are questioned at ACS meeting, *Chemical Week*, v. 152, no. 13, p. 9, 1993.

Rowe, B.L., Landrigan, S.J., and Lopes, T.J., Summary of published aquatic toxicity information and water-quality criteria for selected volatile organic compounds, U.S. Geological Survey Open File Report OFR 97-563, 1997.

Salanitro, J.P., Wisniewski, H.L., and Dortch, I.J., Effects of dissolved oxygen on biodegradation of gasoline components in saturated soil columns (abs.), in Abstracts of the 92nd general meeting of the American Society for Microbiology, American Society for Microbiology, v. 92, p. 354, 1992.

Salanitro, J.P., Diaz, L.A., Williams, M.P., and Wisniewski, H.W., Isolation of a bacterial culture that degrades methyl t-butyl ether, *Applied and Environmental Microbiology*, v. 60, no. 7, p. 2593–2596, 1994.

Salanitro, J., Wisniewski, H., and McAllister, P., Observation on the biodegradation and bioremediation potential of methyl t-butyl ether (abs.), in Society of Environmental Toxicology and Chemistry abstract book, 17th, Washington, D.C., November 17-21, SETAC, p. 115, 1996.

Salanitro, J.P., Chou, Chi-Su, Wisniewski, H.L., and Vipond, T.E., Perspectives on MTBE biodegradation and the potential for in situ aquifer bioremediation, in The Southwest Focused Groundwater Conference—Discussing the Issue of MTBE and Perchlorate in Groundwater, Anaheim, California, June 3-4, 1998, Proceedings, Anaheim, California, National Groundwater Association, p. 40-54, 1998.

Savolainen, H., Pfaffli, P., and Elovaara, E., Biochemical effects of methyl *tertiary*-butyl ether in extended vapor exposure of rats, *Archives of Toxicology*, v. 57, no. 4, p. 285–288, 1985.

Schirmer, M., Barker, J.F., Hubbard, C.E., Church, C.D., Pankow, J.F., and Tratnyek, P.G., The Borden field experiment—Where has the MTBE gone?, in American Chemical Society Division of Environmental Chemistry pre-prints of papers, 213th, San Francisco, California, ACS, v. 37, no. 1, p. 415–417, 1997.

Schirmer, M. and Barker, J.F., A study of long-term MTBE attenuation in the Borden Aquifer, Ontario, Canada, *Groundwater Monitoring & Remediation*, Spring, p. 113-122, 1998.

Schorr, P., Appendix E—Occurrence, treatability and estimated statewide costs to achieve a proposed maximum contaminant level of 70 ppb for methyl *tertiary*-butyl ether in public and non-public drinking water systems in New Jersey, Bureau of Safe Drinking Water, New Jersey Department of Environmental Protection, 1994.

Schroll, R., Bierling, B., Cao, G., Doerfler, U., Lahaniati, M., Langenbach, T., Scheunert, I., and Winkler, R., Uptake pathways of organic chemicals from soil by agricultural plants, *Chemosphere*, v. 28, no. 2, p. 297–303, 1994.

Searle, D.H., Reformulated gasoline—It's supposed to be good for us, *Motorcycle Consumer News*, September, p. 32–35, 1995.

Seunram, R.D., Lovas, F.J., Pereyra, W., Fraser, G.T., and Hight Walker, A.R., Rotational spectra, structure, and electric dipole moments of methyl and ethyl *tert*-butyl ether (MTBE and ETBE), *Journal of Molecular Spectroscopy*, v. 181, no. 1, p. 67–77, 1997.

Sevilla, A., Beaver, P., and Cherry, P., Effect of MTBE on the treatability of petroleum hydrocarbons in water, in American Chemical Society Division of Environmental Chemistry pre-prints of papers, 213th, San Francisco, California, ACS, v. 37, no. 1, p. 403–405, 1997.

Shelley, S. and Fouhy, K., The drive for cleaner-burning fuel, *Chemical Engineering*, v. 101, no. 1, p. 61–63, 1994.

Shen, Y.F., Yoo, L.J., Fitzsimmons, S.R., and Yamamoto, M.K., Threshold odor concentrations of MTBE and other fuel oxygenates, in American Chemical Society Division of Environmental Chemistry pre-prints of papers, 213th, San Francisco, California, ACS, v. 37, no. 1, p. 407–409, 1997.

Smith, D.F., Kleindienst, T.E., Hudgens, E.E., McIver, C.D., and Bufalini, J.J., The photooxidation of methyl *tertiary*-butyl ether, *International Journal of Chemical Kinetics*, v. 23, no. 10, p. 907–924, 1991.

Smith, S.L. and Duffy, L.K., Odor and health complaints with Alaskan gasolines, *Chemical Health & Safety*, v. 2, no. 3, p. 32-38, 1995.

Smylie, M. and Whitten, G.Z., Use of the urban airshed model to generate and evaluate regional-specific reactivity adjustment factors, in Air & Waste Management Association, Annual Meeting & Exhibition, 85th, Kansas City, Missouri, June 21-26, Proceedings, AWMA, no. 92-86.12, 1992.

Sorrell, R.K., Daly, E.M., Weisner, M.J., and Brass, H.J., In-home treatment methods for removing volatile organic chemicals, *Journal of the American Water Works Association*, v. 77, no. 5, p. 72–78, 1985.

Speth, T.F. and Miltner, R.J., Technical note—Adsorption capacity of GAC for synthetic organics, *Journal of the American Water Works Association*, v. 82, no. 2, p. 72–75, 1990.

Squillace, P.J., Pope, D.A., and Price, C.V., Occurrence of the gasoline additive MTBE in shallow groundwater in urban and agricultural areas, U.S. Geological Survey Fact Sheet FS-114-95, 1995.

Squillace, P.J., Zogorski, J.S., Wilber, W.G., and Price, C.V., Preliminary assessment of the occurrence and possible sources of MTBE in groundwater in the United States, 1993-1994, *Environmental Science & Technology*, v. 30, no. 5, p. 1721–1730, 1996.

Squillace, P.J., Pankow, J.F., Korte, N.E., and Zogorski, J.S., Environmental behavior and fate of methyl *tert*-butyl ether (MTBE), U.S. Geological Survey Fact Sheet FS-203-96, 1996.

Squillace, P.J., A review of the environmental behavior and fate of fuel oxygenates (abs.), in Society of Environmental Toxicology and Chemistry abstract book, 17th, Washington, D.C., November 17-21, SETAC, p. 114–115, 1996.

Squillace, P.J., Zogorski, J.S., Wilber, W.G., and Price, C.V., Preliminary assessment of the occurrence and possible sources of MTBE in groundwater in the United States, 1993-1994, in American Chemical Society Division of Environmental Chemistry pre-prints of papers, 213th, San Francisco, California, ACS, v. 37, no. 1, p. 372–374, 1997.

Squillance, P.J., Pankow, J.F., and Zogorski, J.S., Environmental behavior and fate of methyl *tert*-butyl ether (MTBE), in The Southwest Focused Groundwater Conference—Discussing the Issue of MTBE and Perchlorate in Groundwater, Anaheim, California, National Groundwater Association, p. 4–9, 1998.

Stackelberg, P.E., Hopple, J.A., and Kaufman, L.F., Occurrence of nitrate, pesticides, and volatile organic compounds in Kirkwood-Cohansey Aquifer System, Southern New Jersey, U.S. Geological Survey Water-Resources Investigation Report WRIR 97-4241, 1997.

Stackelberg, P.E., O'Brien, A.K., and Terracciano, S.A., Occurrence of MTBE in surface and groundwater, Long Island, New York, and New Jersey, in American Chemical Society Division of Environmental Chemistry pre-prints of papers, 213th, San Francisco, California, ACS, v. 37, no. 1, p. 394–397, 1997.

Stanley, C.C., MTBE — The need for a balanced perspective, in The Southwest Focused Groundwater Conference—Discussing the Issue of MTBE and Perchlorate in Groundwater, Anaheim, California, National Groundwater Association, p. 76, 77, 1998.

Stelljes, M., Issues associated with the toxicological data on MTBE, in American Chemical Society Division of Environmental Chemistry pre-prints of papers, 213th, San Francisco, California, ACS, v. 37, no. 1, p. 401–403, 1997.

Stephenson, R.M., Mutual solubilities—Water-ketones, water-ethers, and water-gasoline-alcohols, *Journal of Chemical Engineering Data*, v. 37, no. 1, p. 80–95, 1992.

Streete, P.J., Ruprah, M., Ramsey, J.D., and Flanagan, R.J., Detection and identification of volatile substances by headspace capillary gas chromatography to aid the diagnosis of acute poisoning, *Analyst*, v. 117, no. 7, p. 1111–1127, 1992.

Stubblefield, W.A., Burnett, S.L., Hockett, J.R., and Naddy, R., Evaluation of the acute and chronic aquatic toxicity of methyl *tertiary*-butyl ether (MTBE), in American Chemical Society Division of Environmental Chemistry pre-prints of papers, 213th, San Francisco, California, ACS, v. 37, no. 1, p. 429–430, 1997.

Stump, F.D., Knapp, K.T., and Ray, W.D., Seasonal impact of blending oxygenated organics with gasoline on motor vehicle tailpipe and evaporative emissions, *Journal of the Air Waste Management Association*, v. 40, no. 6, p. 872–880, 1990.

Suflita, J.M. and Mormile, M.R., Anaerobic biodegradation of known and potential gasoline oxygenates in the terrestrial subsurface, *Environmental Science & Technology*, v. 27, no. 5, p. 976–978, 1993.

Tardiff, R.G. and Stern, B.R., Estimating the risks and safety of methyl *tertiary*-butyl ether (MTBE) and tertiary butyl alcohol (TBA) in tap water for exposures of varying duration, in American Chemical Society Division of Environmental Chemistry pre-prints of papers, 213th, San Francisco, California, ACS, v. 37, no. 1, p. 430–432, 1997.

Task Force on Health Effects of Reformulated Gas, Assessment of the health effects of reformulated gasoline in Maine, Clean Air Stakeholders Conference, The Joint Standing Committee on Natural Resources, Maine State Legislature, May 1995.

Taylor, J.R. and O'Brien, T.J., Evaluating residential water supply wells in a fractured bedrock aquifer contaminated with MTBE—A case study: Groundwater Management, v. 16, p. 929–937, 1993.

Tepper, J.S., Jackson, M.C., McGee, J.K., Costa, D.L., and Graham, J.A., Estimation of respiratory irritancy from inhaled methyl *tertiary*-butyl ether in mice, *Inhalation Toxicology*, v. 6, no. 6, p. 563–569, 1994.

Testa, S.M. and Winegardner, D.L., *Restoration of Contaminated Aquifers, Petroleum Hydrocarbons and Organic Compounds*, Boca Raton, FL, 446 p., 2000.

Truong, K.N. and Parmele, C.S., Cost-effective alternative treatment technologies for reducing the concentrations of methyl *tertiary*-butyl ether and methanol in groundwater, in Calabrese, E.J. and Kostecki, P.T., Eds., *Hydrocarbon Contaminated Soils and Groundwater*, v. 2, Chelsea, Michigan, Lewis Publishers, Inc., p. 461-486, 1992.

Tuazon, E.C., Carter, W.P.L., Aschmann, S.M., and Atkinson, R., Products of the gas-phase reaction of methyl *tert*-butyl ether with the OH radical in the presence of Nox, *International Journal of Chemical Kinetics*, v. 23, no. 11, p. 1003–1015, 1991.

U.S. Environmental Protection Agency, Assessment of potential health risks of gasoline oxygenated with methyl *tertiary*-butyl ether (MTBE), Washington, D.C., Office of Research and Development, EPA/600/R-93/206, 1993a.

U.S. Environmental Protection Agency, An investigation of exposure to MTBE and gasoline among motorists and exposed workers in Albany, New York, Atlanta, Georgia Centers for Disease Control and Prevention, National Center for Environmental Health, Division of Environmental Hazards and Health Effects, and New York State Department of Health, 1993b.

U.S. Environmental Protection Agency, Health risk perspectives on fuel oxygenates, Washington, D.C., Office of Research and Development, EPA 600/R-94/217, 1994.

U.S. Environmental Protection Agency, Drinking water regulations and health advisories, Washington, D.C., Office of Water, EPA 822-R-96-001, 1996.

U.S. Environmental Protection Agency, Drinking water advisory—Consumer acceptability advice and health effects analysis on methyl *tertiary*-butyl ether (MtBE), Washington, D.C., Office of Water, EPA-822-F-97-009, 1997.

U.S. Geological Survey, Denver's urban groundwater quality—Nutrients, pesticides, and volatile organic compounds, USGS Fact Sheet 106-95, 1995.

U.S. National Archives and Records Administration, Federal Registrar's Office, Testing consent order for tertiary amyl methyl ether, Federal Register, v. 60, no. 54, p. 14910-14911, 1995.

Unzelman, G.H., Reformulated gasolines will challenge product-quality maintenance, *Oil & Gas Journal*, v. 88, no. 15, p. 43–48, 1990.

Unzelman, G.H., U.S. Clean Air Act expands role for oxygenates: *Oil & Gas Journal*, v. 89, no. 15, p. 44–49, 1991.

Unzelman, G.H., Impact of oxygenates on petroleum refining—Part I—Historical Review, *Fuel Reformulation*, v. 5, no. 3, p. 51–54, 1995.

Veith, G.D., Call, D.J., and Brooke, L.T., Structure-toxicity relationships for the Fathead Minnow, Pimephales promelas—Narcotic industrial chemicals, *Canadian Journal of Fisheries and Aquatic Science*, v. 40, no. 6, p. 743–748, 1983.

Wallington, T.J., Dagaut, P., Liu, R., and Kurylo, M.J., Gas-phase reactions of hydroxyl radicals with the fuel additives methyl *tert*-butyl ether and tert-butyl alcohol over the temperature range 240-440 K, *Environmental Science & Technology*, v. 22, no. 7, p. 842–844, 1988.

Watson, J.G., Chow, J.C., Pritchett, L.C., Houck, J.A., Ragazzi, R.A., and Burns, S., Chemical source profiles for particulate motor vehicle exhaust under cold and high altitude operating conditions, *The Science of the Total Environment*, v. 93, p. 183–190, 1990.

White, M.C., Johnson, C.A., Ashley, D.L., Buchta, T.M., and Pelletier, D.J., Exposure to methyl *tertiary*-butyl ether from oxygenated gasoline in Stamford, Connecticut, *Archives of Environmental Health*, v. 50, no. 3, p. 183–189, 1995.

Wibowo, A.A.E., DECOS and NEG basis for an occupational standard—Methyl *tert*-butyl ether, Solna, Sverige, National Institute of Occupational Health, 1994.

Widdowson, M.A., Ray, R.P., Reeves, H.W., Aelion, C.M., and Holbrooks, K.D., Investigation of soil venting-based remediation at a UST site in the Appalachian Piedmont, in Schepart, B.S., Ed., *Bioremediation of Pollutants in Soil and Water*, Philadelphia, Pennsylvania, American Society for Testing and Materials, Special Technical Publication 1235, p. 135–148, 1995.

Wiesmann, G. and Cornitius, T., Falling MTBE demand bursts the methanol bubble, *Chemical Week*, v. 156, no. 8, p. 14, 1995.

Williams, C.H., Crow, W.L., and Lewandowski, P.S., Evaluation of community exposure to airborne SARA Title III section 313 chemicals emitted from petroleum refineries, U.S. Environmental Protection Agency's Waste Management Association, Atmospheric Research and Exposure Assessment Laboratory and Waste Management Association, Measurement of toxic and related air pollutants, Journal Code 33847, p. 948–954, 1990.

Williams, B., MTBE, ethanol advocates' squabble may complicate RFG implementation, *Oil & Gas Journal*, v. 93, no. 7, p. 17–22, 1995.

Wilson, E., Scientists wrangle over MTBE controversy, *Chemical & Engineering News*, v. 75, no. 18, p. 54–56, 1997.

Worthington, M.A. and Perez, E.J., Dating gasoline releases using groundwater chemical analyses—Case studies, *Groundwater Management*, v. 17, p. 203–217, 1993.

Xle, Y., and Reckhow, D.A., Formation of halogenated artifacts in brominated, chloraminated, and chlorinated solvents, *Environmental Science & Technology*, v. 28, no. 7, p. 1357–1360, 1994.

Yeh, C.K. and Novak, J.T., Anaerobic biodegradation of oxygenates in the subsurface, in NWWA/API Petroleum Hydrocarbons and Organic Chemicals in Groundwater—Prevention, Detection, and Restoration, Proceedings, Book 8, Houston, Texas, November 20-22, National Water Well Association and American Petroleum Institute, p. 427–441, 1991.

Yeh, K.-J., Degradation of gasoline oxygenates in the subsurface (abs.), Dissertation Abstracts International, v. 53, no. 2, p. 757-B, 1992.

Yeh, C.K. and Novak, J.T., Anaerobic biodegradation of gasoline oxygenates in soils, *Water Environment Research*, v. 66, no. 5, p. 744–752, 1994.

Yeh, C.K. and Novak, J.T., The effect of hydrogen peroxide on the degradation of methyl and ethyl *tert*-butyl ether in soils, *Water Environment Research*, v. 67, no. 5, p. 828–834, 1995.

Yoshikawa, M., Arashidani, K., Katoh, T., Kawamoto, T., and Kodama, Y., Pulmonary elimination of methyl *tertiary*-butyl ether after intraperitoneal administration in mice, *Archives of Toxicology*, v. 68, no. 8, p. 517–519, 1994.

Young, W.F., Horth, H., Crane, R., Ogden, T., and Arnott, M., Taste and odor threshold concentrations of potential potable water contaminants, *Water Research*, v. 30, no. 2, p. 331–340, 1996.

Zhao, X., Smith, S.L., and Duffy, L.K., Effects of ethanol as an additive on odor detection thresholds of Alaskan gasolines at sub-arctic temperatures, *Chemosphere*, v. 31, no. 11/12, p. 4531–4540, 1995.

Zogorski, J.S. et al., Fuel oxygenates and water quality: A summary of current understanding of sources, occurrences in natural waters, environmental behavior, fate and significance, prepared for the Office of Science and Technology Policy, the Executive Office of the President, 1995.

Zogorski, J.S., Fuel oxygenates and water quality—Findings and recommendations of the Interagency Oxygenated Fuel Assessment, in Stanley, A., Ed., NWWA/API Petroleum Hydrocarbons and Organic Chemicals in Groundwater—Prevention, Detection, and Remediation Conference, Proceedings, Houston, Texas, Nov. 13-15, National Water Well Association and American Petroleum Institute, 1996.

Zogorski, J.S., Delzer, G.C., Bender, D.A., Squillance, P.J., and Lopes. T.J., Baehr, A.L., Stackelberg, P.E., Lanmeyer, J.E., Boughton, C.J., Lico, M.S., Pankow, J.F., Johnson, R.L., and Thompson, N.R., MTBE—Summary of findings and research by the U.S. Geological Survey, in 1998 Annual Conference of Water Quality, Atlanta, Georgia, Proceedings, American Water Works Association, 1998.

Index

C

D

E

F

G

M

N

O

P

T

U

V

W